Linking Climate Change Adaptation, Disaster Risk Reduction, and Loss & Damage

Vositha Wijenayake · Linda Anne Stevenson ·
Akio Takemoto · Amit Ranjan ·
Dennis Mombauer · Nafesa Ismail
Editors

Linking Climate Change Adaptation, Disaster Risk Reduction, and Loss & Damage

APN
ASIA-PACIFIC NETWORK FOR
GLOBAL CHANGE RESEARCH

palgrave
macmillan

Editors
Vositha Wijenayake
Executive Director
SLYCAN Trust
Colombo, Sri Lanka

Akio Takemoto
Head of Programme and Administration
UNU Institute for the Advanced Study of Sustainability
Tokyo, Japan

Dennis Mombauer
Director, Research & Knowledge Management
SLYCAN Trust
Colombo, Sri Lanka

Linda Anne Stevenson
Programme Director
Asia-Pacific Network for Global Change Research (APN)
Kobe, Japan

Amit Ranjan
Institute of South Asian Studies
National University of Singapore
Singapore, Singapore

Nafesa Ismail
Senior Programme Officer
Asia-Pacific Network for Global Change Research (APN)
Kobe, Japan

ISBN 978-981-99-8054-3 ISBN 978-981-99-8055-0 (eBook)
https://doi.org/10.1007/978-981-99-8055-0

This Palgrave Macmillan imprint is published by the registered company Springer Nature Singapore Pte Ltd.
The registered company address is: 152 Beach Road, #21-01/04 Gateway East, Singapore 189721, Singapore

Paper in this product is recyclable.

We dedicate this book to Professor Saleemul Huq, who left the world on 28 October 2023. Saleem's most notable legacy lies in his dedication to addressing climate-induced loss and damage. He passionately championed this cause, ultimately ensuring the inclusion of loss and damage in the Paris Agreement in 2015, and the establishment of a finance mechanism for loss and damage at COP27 in 2022. Saleem has written the foreword to this book.

Foreword

Climate change is one of the most pressing challenges of our time, and its impacts are becoming increasingly visible and alarming. The urgency to act on available scientific evidence and generate more evidence on these impacts, particularly loss and damage, cannot be overstated. This publication aims to delve into evidence-based approaches in the Asia-Pacific region to effectively tackle climate-induced loss and damage (L&D) and its interlinkages with climate change adaptation (CCA) and disaster risk reduction (DRR) in the context of the Sustainable Development Goals (SDGs) and other global initiatives and frameworks.

The Sixth Assessment Report (AR6) recently published by the Intergovernmental Panel on Climate Change (IPCC) provides a comprehensive assessment of our current understanding of climate change, including the topics of L&D and CCA. The report highlights various types of climate-induced L&D, ranging from extreme weather events like floods, cyclones and wildfires, to slow-onset impacts such as sea-level rise, ocean acidification and desertification. As indicated by climate science and evidence from the ground level, the impacts of climate change are expected to worsen in the future, particularly in areas with high vulnerability and exposure. It is essential to recognize that different countries face these impacts differently, underscoring the need to build resilience, capacities and means of implementation.

Furthermore, it is crucial to acknowledge the interconnections between national and global policy processes, including those under the United Nations Framework Convention on Climate Change (UNFCCC), such as Nationally Determined Contributions (NDCs) and various other plans and policies relevant for addressing climate-induced L&D. A holistic approach to addressing these challenges necessitates leveraging synergies between different frameworks and initiatives. The SDGs and the Sendai Framework for Disaster Risk Reduction (SFDRR) provides avenues for entry points for creating scaled-up impact related to L&D policy and action, by building on national, regional and global actions through forging initiatives built on commonalities. By engaging multiple stakeholders and embracing inclusive and participatory approaches, we can drive transformative change at the local, national and global levels.

Scientific evidence must play a pivotal role in guiding our actions to effectively address L&D. It is imperative to emphasize science-based approaches and leverage evidence, research and capacity building to accelerate transformative action. In this regard, this publication by the Asia-Pacific Network for Global Change Research (APN) in partnership with SLYCAN Trust contributes to generating evidence that highlights the power of scientific knowledge generation involving government entities, civil society organizations, academia and multilateral institutions.

Looking ahead, this publication aims to serve as a catalyst for change. Through the dissemination of knowledge and the facilitation of a stronger science-policy-stakeholder dialogue, we aspire to create tangible and positive impacts on the ground. Effectively addressing L&D in the Asia-Pacific region necessitates collective efforts and synergies among diverse processes. It is through integrated and collaborative endeavours that we can achieve scaled-up impacts, empower communities and safeguard our shared future.

To achieve this transformation and uphold efforts for resilience-building that leaves no one behind, it is crucial to address key aspects such as building capacities, bridging gaps and identifying policy interventions. By acknowledging and acting upon these imperatives, a path towards resilience-building could be forged through collective efforts that deliver transformative change. By integrating the principles of leaving no one behind, climate change policy and action could prioritize inclusivity, equity and justice, empowering vulnerable communities to thrive amidst environmental challenges and creating an enabling environment that fosters sustainable development leading to an equitable future.

Saleemul Huq
ICCCAD
Bangladesh

Saleemul Huq was the Director of the International Centre for Climate Change and Development (ICCCAD) and professor at the Independent University Bangladesh (IUB), as well as a senior associate of the International Institute on Environment and Development (IIED) in the United Kingdom. In addition, he was the chair of the Expert Advisory Group for the Climate Vulnerable Forum (CVF) and also a senior adviser on Locally Led Adaptation with Global Centre on Adaptation (GCA) headquartered in the Netherlands. He was an expert in adaptation to climate change in the most vulnerable developing countries and a lead author of the third, fourth, and fifth assessment reports of the Intergovernmental Panel on Climate Change (IPCC) and he also advised the Least Developed Countries (LDC) group in the United Nations Framework Convention on Climate Change (UNFCCC). In addition, he was affiliated with the UN Food System Summit for 2021 as co-chair of the Action Track 5 on Building Resilience to Vulnerabilities, Shocks & Stress. He published hundreds of scientific as well as popular articles and was recognized as one of the top twenty global influencers on climate change policy in 2019 and top scientist from Bangladesh on climate change science. Recently, he had been appointed the officer of the Order of the British Empire (OBE) in the 2022 New Year Honours for services to combating international climate change. In 2023, he had been ranked as one of the top ten scientists in Nature, one of the leading scientific journals.

Contents

Editors and Contributors

About the Editors

Vositha Wijenayake works as an Executive Director of SLYCAN Trust and as a researcher and attorney-at-law specializing in environmental law and public international law. She has over a decade of experience in working on climate change at the national and international levels, focusing on issues related to adaptation, Nationally Determined Contributions (NDCs), National Adaptation Plans (NAPs), loss and damage, human mobility, resilience, climate impacts, governance, vulnerable communities, gender, ecosystems, sustainable consumption and production, and sustainable development.

Linda Anne Stevenson is Programme Director at the Asia-Pacific Network for Global Change Research (APN), an Intergovernmental network of 22 member countries. With over 22 years of experience in APN, she oversees the management of over eighty national and regional research and capacity building projects across a broad range of disciplines in the global environmental research domain and works closely with researchers and decision-makers in the Asia-Pacific region and beyond. She has significant experience in scientific capacity building, including developing early career professionals' capacity; planning dialogues that cross the science-policy-practitioner interface; and engaging with international organizations and bodies, including UNFCCC, IPCC, and IPBES.

Dr Stevenson was also an expert reviewer for the Working Group II Report of the 6th Assessment Report of the IPCC and for a number of IPBES assessment reports.

Akio Takemoto is a Programme Head at the United Nations University Institute for the Advanced Study of Sustainability (UNU-IAS) based in Tokyo, Japan. He leads and coordinates academic and administrative activities of UNU-IAS projects in the institute's three thematic areas: sustainable societies, natural capital and biodiversity, and global change and resilience. From January 2018 to January 2020, Dr. Takemoto served as a senior environmental specialist at Global Environmental Facility (GEF). He managed the GEF programmes in Climate Change and Chemicals & Waste focal areas at the GEF Secretariat in Washington, DC, USA. From July 2017 to January 2018, Dr. Takemoto served in climate change negotiation as the Director for International Strategy on Climate Change at the Ministry of the Environment, Japan (MOEJ). From July 2014 to July 2017, he served as the director of the Research and Information Office and Climate Change Adaptation Office at MOEJ.

Amit Ranjan is a research fellow at the Institute of South Asian Studies, National University of Singapore. His latest books include the following: (with Rajesh Kharat and Pallavi Deka) Environment, Climate Change and Migration in South Asia (Routledge, Oxon, 2023), (with Ian Talbot) Urban Development and Environmental History in Modern South Asia (Routledge, Oxon, 2023), and Contested Waters: India's Transboundary River Water Disputes in South Asia (Routledge, Oxon and New Delhi, 2021). He is also the author of India-Bangladesh Border Disputes: History and Post-LBA Dynamics (Springer, Singapore, 2018). He has edited the following books: India in South Asia Challenges and Management (Springer, Singapore, 2019), Partition of India: Post-colonial Legacies (Routledge, Oxon and New Delhi, 2019), and Water Issues in Himalayan South Asia: Internal Challenges, Disputes and Transboundary Tensions (Singapore, 2019). His paper review essays and book reviews have been widely published in journals, including Asian Survey, Asian Affairs, Asian Ethnicity, Asian Journal of Comparative Politics, Economic & Political Weekly, India Review, Indian Journal of Public Administration, India Quarterly, Journal of Migration Affairs, Journal of the Indian Ocean Region, Roundtable: The Commonwealth Journal of International Affairs, Social Change, Studies in Indian Politics, Society and Culture in South Asia, South Asia Research, Journal of Asian Security and International Affairs, Water History, Journal of African and Asian Studies, and World Water Policy.

Dennis Mombauer works as a Director of Research & Knowledge Management at SLYCAN Trust and has over a decade of experience in research, management, administration, and coordination in the private sector and the development sector. His work focuses on thematic areas related to climate change, adaptation and resilience, loss and damage, climate finance, risk management, just transition, sustainable development, human mobility, ecosystem conservation and restoration, and other related topics.

Nafesa Ismail has been working at the Asia-Pacific Network for Global Change Research (APN) for over five years in the capacity as Senior Programme Officer. She supports the management of national and regional projects in the Asia-Pacific region and the management and monitoring of the knowledge generated from APN activities, including the APN Science Bulletin, a flagship web-based and peer-reviewed publication. She also supports the empowerment of early career professionals in leading and participating in international research through the APN's proposal development training workshop. Dr Ismail received her doctorate from Kyoto University and a Master's degree from Kansai University.

Contributors

Dannica Rose G. Aquino is a Project Technical Specialist I for the DOST-PCAARRD project titled 'Natural Capital Accounting of Coastal and Marine Ecosystems in the West Philippine Sea (WPS).' With a background in environmental science, Dannica brings expertise in assessing the impacts of interventions in agriculture, forestry, and natural resource sectors. Her previous engagements have fueled her research interests in natural resource valuation and policy analysis. Dannica holds a Master's degree in Environmental Science from the University of the Philippines Los Baños.

Senaka Basnayake works as a Director for the Climate Resilience Department at Asian Disaster Preparedness Center. He holds a Ph.D. Degree in meteorology and has more than 30 years of experience in project management and strategic guidance for implementing climate risk management, climate change adaptation, urban resilience projects and programmes in Asia and the Pacific. He is experienced in conducting research on severe weather events, understanding climate change and variability and has extensive experience in leading the development of climate services, climate modelling, impact and vulnerability assessments for managing climate risks. In recent years, he has been involved in estimation of economic and non-economic losses and damages to various types of hazards to support climate change adaptation in vulnerable sectors.

Rizaldi Boer is an agroclimatologist and a senior researcher at the Centre for Climate Risk and Opportunity Management of IPB University. His research focuses on climate change mitigation and adaptation, particularly on agriculture, forest, and other land uses. He is very active in conducting several studies related to policies on low carbon and climate resilience development. He is a member of the Asia Low Carbon Research Network (LoCARNet) and the Indonesian Agriculture Meteorology Society.

Michael Boyland is a Ph.D. candidate in Development Studies at SOAS, University of London. His research concerns how young people are engaging in climate change adaptation and activism in Southeast Asia. Michael was previously a research fellow at the Stockholm Environment Institute (SEI) Asia Centre, working to better understand climate change, disasters and development processes in the region. He has co-authored more than 20 peer-reviewed publications on a range of climate, disasters, and development issues with scholars and practitioners across Asia.

Yohei Chiba works as a research fellow at Disaster Resilience Research Division, National Research Institute for Earth Science and Disaster Resilience (NIED) in Japan. His interests include public policy, finance, accounting, and climate change adaptation. He has also been conducting research in the fields of disaster resilience finance and business continuity of companies.

Richard T. Cooper is an experienced environmental scientist and an international project manager with over 25 years of experience contributing to environmental, sustainability, and climate change-related initiatives in Asia, the Middle East, Africa, Central America, Australia, and the United Kingdom. His interests include decision support systems, environmental informatics, geographic information systems, climate change adaptation, and marine and coastal management. He holds a Ph.D. in Environmental Studies from Macquarie University, Australia, and an M.Sc. in Marine and Fisheries Science from Aberdeen University, UK. Currently, he works as a senior environmental specialist with the International Centre for Environmental Management (ICEM) in Vietnam.

Kees van der Geest Ph.D. is the Head of the Environment and Migration: Interactions and Choices (EMIC) Division at the United Nations University Institute for Environment and Human Security (UNU-EHS). As a human geographer, he studies the impacts of climate change, human mobility, adaptation, and rural livelihoods. Key features of his work are the people-centred perspective and the mixed-method approach combining quantitative and qualitative research tools. His work has contributed substantially to expanding the empirical evidence base on migration-environment linkages and impacts of climate change beyond adaptation ("loss and damage"). Kees has extensive fieldwork experience in the Global South, mostly in Ghana (5 years), but also in Burkina Faso, Vietnam, Bangladesh, Nepal, and the Marshall Islands.

Niladri Gupta is a Senior Water Resources Management Specialist at Asian Disaster Preparedness Center, Thailand. He has a Ph.D. in Geography and Environment with a focus on Earth Observation and Fluvial Geomorphology. His current work focuses on climate adaptive water resources management in Asia and the Pacific region. He also has an interest in the application of Nature-based Solutions for management of water resources to build resilience to climate change. He is currently leading the water sector portfolio in a World Bank-funded project on Climate Adaptation and Resilience for South Asia.

Hop T. B. Hoang is an early career researcher from Vietnam. She has a strong commitment to environmental protection and sustainable development. Her work focuses on climate change mitigation and adaptation, disaster risk management, and natural resources management, especially in public policy and agricultural sectors. She is currently a member of Young Networks working on the mentioned sectors, and she is looking for opportunities for further contribution to these fields.

Cao Thi Thu Huong is an expert in the fields of environment, climate change adaptation, measurement, reporting, and verification for both technical and policy aspects with 20 years of professional experience. She has been a climate change specialist for numerous urban development projects as well as preparing green growth and climate change action plans for the transport sector funded by the Vietnam government and international donors. Ms. Huong holds a Masters of Environmental Science from the Vietnam National University in 2003 and Executive Masters in Development Policies and Practices from The Graduate Institute Geneva, Switzerland, in 2020.

Rotchana Intharathirat is a senior professional environmentalist and director of the Division of Solid Waste and Hazardous Substances Management, Environment and Pollution Control Office 9, Pollution Control Department, Ministry of Natural Resources and Environment, Udon Thani, Thailand. Dr. Rotchana has more than 25 years of experience in research, consultancy, and capacity building in the areas of environment, climate change mitigation, waste to energy, and carbon footprints for organizations. She has published two book chapters and four refereed journal articles and conference papers. She has two licences, professional environmental engineer and associate civil engineer of the Council of Engineers and the Engineering Institute of Thailand.

Md. Atikul Islam works as a Professor at the Environmental Science Discipline, Khulna University. He obtained his Ph.D. in Doctor Engineering in Systems Design and Engineering from the Graduate School of Science and Engineering, Yamaguchi University, Japan, in 2008. His research interests are environmental pollution and public health.

Susantha Jayasinghe is a Senior Climate Data Specialist at the Asian Disaster Preparedness Center, Thailand. His academic expertise is in meteorology, applied geoinformation, and information and communication technology (ICT). His professional background is in meteorological and hydrological modelling, forecasting and early warning systems, including data processing, numerical weather prediction, post-processing, analysis and visualization, access of data from various sources, hardware and software configuration and installation, implementation of forecasting and early warning systems in National Meteorological and Hydrological Services. His current research focuses on the use of geospatial data and digital intelligence development for local-level decision-making.

Ham Kimkong is a senior research fellow and a lecturer at the Department of Natural Resource Management and Development, Faculty of Development Studies, Royal University of Phnom Penh (RUPP) in Cambodia. He is also a final-year Ph.D. candidate at the Faculty of

Humanity and Social Science, Khon Kaen University. Mr. Kimkong has been involved in a variety of national and international agencies working on water governance, agrarian transformations, and disaster recovery narratives. In addition, he runs a national consultancy firm working on Environmental and Social Impact Assessment, mining, special economic zones, and water resource management projects.

Rodel D. Lasco has more than 40 years of experience in natural resources and environmental research, conservation, education and development at the national and international levels. His work has focused on issues related to natural resources conservation, climate change, and land degradation. He is an author of several reports of the IPCC, the 2007 co-winner of the Nobel Peace Prize. He is also a member of the National Academy of Science and Technology (NAST) in the Philippines. He is the executive director of the Oscar M Lopez Center, a private foundation whose mission is to promote action research on climate adaptation and disaster risk reduction.

Jaranporn Lertsahakul is the CEO of Green Style Co., Ltd. Bangkok, Thailand. She has over ten years of experience in research, consultancy in the areas of environment, climate change mitigation and carbon footprints for organizations. She is a national committee member of the Green Office Campaign under the Department of Climate Change and Environment, Ministry of Natural Resources and Environment in Thailand. She has consulted for many organizations of certificated Carbon Footprint for Organizations and helped to obtain carbon credits through the Thailand Voluntary Emission Reduction Project, assisting them in achieving their goal as Carbon Neutral Organizations.

Hina Lotia is a Climate and Development expert with 23+ years of technical experience in research and public policy engagement. She has worked with governments, development partners, and the non-profit sector across Asia and the Pacific. She led the project titled Methods Toolbox for Assessing Loss and Damage at Local Level, publications from which were cited in the WGI and WGII contribution to the IPCC Sixth Assessment Report (AR6). Won the Annual APPAM/JCPA Best Comparative Policy Paper Award 2019 by the Association for Public Policy Analysis and Management and the Journal of Comparative Policy Analysis. Member of Alliance for Global Water Adaptation (AGWA)

Strategic Advisory Council, Board Member of Water Environment Forum (WEForum), Co-coordinator of Upper Indus Basin Network (Pakistan).

Danny Marks Ph.D. is an Assistant Professor of Environmental Politics and Policy in the School of Law and Government of Dublin City University. Dr. Marks has spent a number of years conducting research and working in Southeast Asia, particularly in the field of environmental governance. He has published a number of articles on political ecology, environmental justice, climate governance, and disaster risk reduction, with a focus on Southeast Asia. He completed his Ph.D. dissertation, An Urban Political Ecology of the 2011 Bangkok Floods, at the University of Sydney.

Tran Quang Minh holds a Master's degree in Hydrology and Environment. He has worked at the Sub-Institute of HydroMeteorology and Climate Change (SIHYMECC) in Vietnam, and the Mekong River Commission Secretariat (MRCs) in Lao PDR and Cambodia. He has almost 38 years of experience, including as a project leader and researcher in international, national, and ministerial projects that address flooding, salinity intrusion, climate change, environment from 1985 to the present.

Nguyen Dang Mau currently serves as the Acting Director of the Research Center for Agricultural Meteorology (RCAgM) at the Vietnam Institute of Meteorology, Hydrology, and Climate Change (IMHEN). With a focus on interdisciplinary climate research, Dr Mau delves into socio-economic impacts, urban climate, community health, AI research, climate modeling, and its ramifications.

Pham Thi Thanh Nga is the Director General of the Vietnam Institute of Meteorology, Hydrology, and Climate Change (IMHEN). She is a leading expert in climate, meteorology, and remote sensing. Specializing in disaster assessment and environmental monitoring, Assoc. Prof. Nga adeptly utilizes remote sensing technologies, such as satellites and radar, in conjunction with Geographic Information Systems (GIS) to analyse natural disasters associated with weather and climate. Her extensive experience includes leading various government-funded projects, particularly in meteorological satellite applications, weather forecasting, and disaster damage assessment using remote sensing data. With a proven track record of successful collaboration with international colleagues and specialists, Assoc. Prof. Nga's expertise in the nexus of remote sensing, natural disasters, climate change, and meteorology uniquely positions her to contribute significantly to the advancement of the institute.

Ninh H. Nguyen is a Chairman of the Center for Environment Research, Education and Development (CERED) based in Vietnam. He is a professor, doctor (Hon.) at the University of Pécs (Hungary), an adjunct professor at San Diego State University (USA), an honorary professor at Nguyen Tat Thanh University (Vietnam), and a doctor of Science (Hon.) of the University of East Anglia (UK). He was a contributor to the 4th Assessment Report of the Intergovernmental Panel on Climate Change (IPCC) when it received a Nobel Peace Prize in 2007. He focuses on natural resource management, climate change adaptation and mitigation, disaster risk management, and sustainable development and supports public-private partnerships.

Le Minh Nhat is currently a Deputy Director of the Department of Natural Disaster Response and Recovery, Vietnam Disaster Management Authority (VNDMA), Ministry of Agriculture and Rural Development (MARD). He has more than 20 years of practical experience in Disaster Risk Reduction (DRR) and Management in Vietnam, leading the development of policies and strategies for DRR. Previously, he was the director of the Climate Change Adaptation Division (CCA Vietnam) of the Department of Meteorology, Hydrology and Climate Change (DMHCC), Ministry of Natural Resources and Environment Vietnam (MONRE). He completed his Ph.D. in Hydrology at Kyoto University, Japan, in 2008.

Lam Vu Thanh Noi is currently a vice manager of the Training and International Cooperation Department and a senior researcher of SIWRR. She received her Ph.D. degree from the Asian Institute of Technology (AIT), Bangkok, Thailand. She has 20+ years of experience and worked on water supply planning, water policy and governance for water security, water demand management and public engagement, water resource management, climate change risk assessment and adaptation for the water sector in Southeast Asia. Dr. Lam Noi was a principal investigator of APN-funded project "Climate change risk assessment and adaptation for loss and damage of urban transportation infrastructure in Southeast Asia". She is also a lecturer with teaching experiences at AIT and Dalat University. Currently, Dr. Lam is a principal investigator of a Stockholm Environment Institute-funded project (SUMERNET 4ALL) with study areas in Lao PDR, Thailand, and Vietnam. She is also the team leader of some international and national research projects at SIWRR that focus on water resources policies and assessment, water security, and climate change adaptation for water resources management in Vietnam. She has been involved in several projects funded by international donors and organizations, including AIT, APN, ADB, BMUB, SIDA, and the World Bank.

Agus Nugroho is a programme manager at the Stockholm Environment Institute (SEI) Asia Centre and for the Sustainable Mekong Research Network (SUMERNET). He has over 15 years of experience in research, project management, monitoring and evaluation, and network development in the Mekong and Southeast Asia. Agus has managed and coordinated several research and capacity building programmes in various international and regional organizations, including the ASEAN Secretariat, GTZ, UNDP, and SEI. Agus holds a dual international Master's degree in project management and management of technology.

Chhinh Nyda is a lecturer at the Royal University of Phnom Penh. He completed his Ph.D. at the School of Environment, Flinders University, Australia. Chhinh's research interests include climate change adaptation and rural development.

Joy Jacqueline Pereira is a vice-chair of Working Group II of the IPCC for the Sixth Assessment Report. Previously, she has served as a coordinating lead author for Chapter 24 on Asia of the IPCC Fifth Assessment Report (AR5), lead author for the IPCC AR5 Synthesis Report, and a review editor for the 2012 IPCC Special Report on Managing the Risks of Extreme Events and Disasters to Advance Climate Change Adaptation (IPCC SREX).

Sivapuram Venkata Rama Krishna Prabhakar works as a Principal Policy researcher at The Institute for Global Environmental Strategies (IGES), Japan. His research interests are vulnerability assessments, transboundary climate risks, loss and damage, mainstreaming CCA and DRR into development plans and policies, risk insurance, adaptation metrics, adaptive policies and institutions, loss and damage, and capacity needs assessments using mixed-methods approaches.

Juan M. Pulhin is a full Professor and former Dean of the College of Forestry and Natural Resources, University of the Philippines Los Baños

(UPLB), Philippines. He also holds the highest rank of UP Scientist III since 2011. He was a four-time Visiting Professor at the University of Tokyo and a Visiting Scholar at the Waseda University, Tokyo, Japan.

Muhammad Habibur Rahman is an associate professor of Economics at Durham University UK and Curtin University Australia. Prior to Durham, he held research appointments at Deakin University, the University of Melbourne, and Monash University Malaysia. Before moving to academia, he worked for several international development partners, such as the United Nations Development Programme (UNDP) and Asian Disaster Preparedness Centre (ADPC) in Thailand. His research interests are at the crossroads of Development Economics and Environmental Economics. Recently, he has conducted experiments related to the economics of exogenous shocks (e.g., natural disasters, climate change, and famines) and their bearing on economic performance and political institutions. He has been successful in attracting several external grants from the Australian Government's Department of Foreign Affairs and Trade (DFAT), the Royal Norwegian Foreign Affairs, the Global Facility for Disaster Risk Reduction (GFDRR), the Asia-Pacific Network for Global Change Research (APN), the Ministry of Higher Education Malaysia, and Bushfire & Natural Hazard Cooperative Research Centre (BNH-CRC) of Australia. He regularly extends his technical assistance to government and non-government organizations. He has published in well-recognized academic journals such as the Journal of Development Economics, Journal of Economic Behavior & Organization, Health Economics, Energy Economics, Economic Record, Southern Economic Journal, Public Choice, and Economic Modelling.

Sindra Sharma is the Global Policy Lead at Climate Action Network International. She has a Ph.D. in Environmental Policy and International Development from the London School of Economics.

Harjeet Singh is a global expert on the issues of climate impacts, migration, and adaptation. He has supported countries across the world on tackling climate change and coordinating emergency response and disaster resilience programmes. He is the head of Global Political Strategy at Climate Action Network International. He is a member of the United Nations Technical Expert Group on Comprehensive Risk Management under the Warsaw International Mechanism for Loss and Damage.

Sath Sitak holds a Master's in Urban Planning from the University of Melbourne (Australia), awarded in 2021, and a Bachelor of Environmental Science from the Royal University of Phnom Penh in 2015. He has worked with the Ministry of Environment since 2015 on green or sustainable city development, green building, strategic environmental assessment (SEA), ecolabelling, project development and management, and research. His areas of interest include urban spatial planning and climate change, environmental assessment, and the sustainable built environment.

Nguyen Van Thang Dr. Thang is the Chief Editor of the Journal of Climate Change Science. He completed his Doctor of Philosophy in Meteorology at Lomonosov Moscow State University, Moscow. He is a distinguished member of several academic societies and has contributed extensively to meteorology, climate change, training, and international cooperation.

Dang Quang Thinh is currently the Director of the Climate Change Research Centre under the Vietnam Institute of Meteorology, Hydrology and Climate Change. He completed his Ph.D. at the University of Augsburg in Geography. His research interests include flooding simulations, drought analysis, integrated water resource management, disaster risk reduction, climate change adaptation and mitigation, and loss and damage.

Tran Thi Tinh is a lecturer at Da Lat University. Dr. Tinh has 23 years of teaching and research experience in aquatic ecosystems and sustainable management of natural resources. She has implemented projects such as Application AQUATOX models in managing water quality and studied the impact of climate change on phytoplankton composition in reservoirs.

Tran Thanh Thuy is the Director of the Department of Science, Training, International Cooperation and Journal at the Vietnam Institute of Meteorology, Hydrology, and Climate Change (IMHEN). She holds a PhD in Climate Change from IMHEN, and specializes in enhancing national capacities to respond to climate change in Vietnam. Her research promotes human rights and gender equality through climate change actions and disaster risk management. Additionally, she contributes to identifying areas where climate change action aligns with broader government policy targets.

Lan Anh Tong research interest includes the topics of development economics investigating the effects of policy/programmes on economic development outcomes, such as agricultural productivity, agricultural resources, international trade, education, and health. Her research also provides insights into the socio-economic consequences of natural disasters on the communities and how disaster relief programmes aid community recovery, especially the vulnerable groups. Tong research has been published in several well-recognized economic journals, such as the American Journal of Agricultural Economics, the Australian Journal of Agricultural and Resource Economics, and Economic Record.

Mehmet Ali Ulubasoglu is a professor of Economics and the director of the Centre for Disaster Resilience and Recovery at Deakin University, Australia. He is a leading Australian expert on the economics of natural disasters, with a particular focus on the impact of natural disasters in Australia and Asia on household income, labour markets, and economic sectors, as well as on the health and mental well-being of economically and sociologically vulnerable groups. He has also published several articles on disasters and political development in cross-national contexts. He has taught several courses in the undergraduate, Ph.D., and Executive M.B.A. programmes at Deakin and supervised 15 Ph.D. students as a doctoral supervisor. Ulubasoglu has also undertaken several research projects with government organizations both nationally and internationally and held consultancy roles for the World Bank and the Asian Development Bank.

Spoann Vin obtained his Ph.D. from Okayama University (Japan) and M.Sc. in Urban Environmental Management from the Asian Institute of Technology in 2005. Most of his research projects are related to safeguards for infrastructure management and implementation. For years of his professional background and expertise, he has led and been involved in various projects funded by UNDP, ADB, WB, WWF, AIT-CIDA, IFAD, and EIB. He has knowledge of evaluating and assessing development/investment projects. His recent research focuses on the assessment of climate change risk and adaptation for loss and damage of transportation infrastructure in Battambang province of Cambodia.

List of Figures

Chapter 9

Chapter 10

Chapter 11

Chapter 13

List of Tables

Chapter 7

Chapter 9

Chapter 10

Chapter 11

Chapter 12

Chapter 13

CHAPTER 1

Introduction

Vositha Wijenayake

Abstract Climate change has become one of the key issues the planet is facing today. With impacts being felt across the world, identifying ways to build long-term resilience to its impacts has become a crucial component for a sustainable and resilient planet. This chapter discusses the approaches for ensuring that synergies and interlinkages across key global and national processes are generated to ensure that inclusive and collective actions can be generated through community-centred, evidence-based, inclusive and multi-stakeholder processes. With case studies and examples from projects conducted in the Pacific region, the chapter demonstrates the importance of integrating processes and relevant mechanisms on climate change adaptation, disaster risk reduction, climate-induced loss and damage, and sustainable development and builds on the evidence provided through insights from the case studies to provide recommendations for holistic policy and action for enhanced impact through forward-thinking synergized actions across processes.

V. Wijenayake (✉)
SLYCAN Trust, Colombo, Sri Lanka
e-mail: vositha@slycantrust.org

V. Wijenayake et al. (eds.), *Linking Climate Change Adaptation, Disaster Risk Reduction, and Loss & Damage*,
https://doi.org/10.1007/978-981-99-8055-0_1

Keywords Climate change · SDGs · Sustainable development · Disaster risk reduction · Sendai Framework · Cross-cutting

1 Background

Climate change is amongst the key threats to our planet. The Sixth Assessment Report of the Intergovernmental Panel on Climate Change (IPCC, 2022) points to the heightened need for climate adaptation and resilience-building as well as interlinkages between addressing climate risks and building the long-term resilience of communities and ecosystems. To this end, it is important to identify through evidence-based approaches the initiatives for building synergies between climate change adaptation (CCA), addressing climate-induced loss and damage (L&D), disaster risk reduction (DRR), sustainable development, and other relevant processes. Evidence-based policy and action, which facilitate good governance through inclusive and transparent processes could provide the core for ensuring that long-term climate resilience is achieved, and scaled-up with initiatives that draw synergies with global and regional processes.

2 Objectives

This publication, a joint effort of the Asia–Pacific Network for Global Change Research (APN) and SLYCAN Trust, aims to demonstrate, through research and evidence from projects conducted in countries of the Asia-Pacific region, initiatives and good practices that could provide opportunities for generating approaches that lead towards building long-term and sustained resilience at all levels. It consists of key findings of projects conducted under APN's Climate Adaptation Framework (CAF) and reflects the work of fourteen projects concentrated on the thematic areas of CCA, DRR, and L&D.

Further, the publication illustrates concrete efforts for evidence-provision that could provide support in identifying solutions to addressing identified gaps and needs; replicate good practices; provide a knowledge base for generating science-driven climate policy and action in the Asia–Pacific region; as well as share initiatives that may be replicated for scaling up impact for building long-term resilience.

3 Overview

3.1 Focus

The publication has as its focus the interlink between CCA, DRR, L&D, and the Sustainable Development Goals (SDGs) through different components related to the thematic areas. This includes through a broader focus, global policy processes related to climate change, DRR, and sustainable development, which in turn include the processes on the United Nations Framework Convention on Climate Change (UNFCCC); the Sendai Framework for Disaster Risk Reduction (SFDRR); the Agenda 2030 for Sustainable Development; and other national and regional initiatives that interlink with the global climate policy and action.

Further, the chapters of the publication provide insights into emerging issues such as the means of implementation and needs related to addressing climate-induced L&D, and forward-thinking approaches building on the findings of the featured papers as well as the most recent climate science.

This special issue could also be described as an effort to demonstrate the scalability of initiatives that integrate research efforts focused on theoretical and technical aspects related to national and global policy processes; as well as research focusing on local-level approaches for resilience-building and addressing climate and disaster risk management through inclusive and participatory processes. The quality and the richness of the evidence generated through the projects featured can be considered as amplified through the reference to their good practices via citations made to these research and related publications in the IPCC's Sixth Assessment Report. This publication aims to ensure that the good practices and lessons learnt through the projects reach a wider audience with enhanced accessibility.

3.2 Thematic Overview

The publication consists of chapters that illustrate different aspects related to interlinkages and integration of CCA, DRR, L&D, and SDGs, which guide the delivery of holistic and scaled-up climate-resilience-building efforts. The thirteen chapters of the publication could be categorized as delivering key insights into elements which connect the different focus areas mentioned above.

This includes research focused on identifying and defining the interlinkages between CCA, DRR, and L&D; means for assessing climate risk which form common grounds for building resilience through reduction of risk; identifying and breaking down of different forms of L&D including but not limited to economic and non-economic L&D, as well as climate-induced migration; and interlinks between DRR and L&D, which include approaches for addressing losses and damage through recovery processes. In addition to these, the publication also focuses on approaches for addressing climate risk through futuristic and forward-thinking approaches, which consist of identifying key entry points, policy and action processes, and approaches for creating synergies.

3.3 Interlinkages Amongst Processes Related to CCA, DRR, L&D, and SDGs

Amongst the chapters of the publication, several provide examples and findings related to interlinkages and the creation of synergies amongst the above-mentioned processes. For example, the chapter "Flood induced losses and damages to agricultural production and farmers' mental well-being in Sri Lanka," (Basnayake et al., 2023) focuses on interlinkages of climate impacts, L&D, economic resilience-building, and empowerment with a central focus on food security. Further, the chapter also looks into future projections with regard to losses and damages, based on the data gathered as part of the relevant research.

The chapter "SWOC analysis on the proposed linkage between Climate Change Adaptation (CCA), Disaster Risk Reduction (DRR), and Loss and Damage (L&D): Case Studies in the low-lying coastal cities of Indonesia, Philippines, Thailand and Vietnam" (Hoang et al., 2023) focuses on the interlinkages of the three processes. This includes components which focus on interlinkages across the processes in Indonesia, the Philippines, Thailand, and Vietnam with a focus on coastal sector, and coastal cities. The chapter also points to the need for better integration of different interrelated processes for heightened impact. The chapter also highlights the need for multi-stakeholder engagement and contribution for reducing vulnerability to climate risk and impacts.

3.3.1 Addressing Interlinkages Through Policy Approaches

Expanding the conceptualization of interlinkages building amongst the highlighted thematic areas, the chapter "Linkages between disaster risk reduction and climate change adaptation in the context of increasing climate change-induced loss and damage in Vietnam" (Nhat and Thinh, 2023) studies the interlinkages amongst different thematic areas through a national lens. Focused on Vietnam, the chapter aims to identify interlinkages at a legal and policy level. In addition, the chapter also seeks to identify the appropriate allocation of efforts across the several thematic areas for generating better results and impacts.

The chapter "Policy gaps and needs analysis for the implementation of NDCs on adaptation and loss and damage in Bangladesh, Nepal, and Sri Lanka," (Mombauer and Wijenayake, 2023) provides a similar approach in identifying interlinkages amongst the areas of CCA and L&D but through a broader approach which focus on three countries. The chapter further maps the interlinkages based on the NDCs under the Paris Agreement, with an aim to succeed in building regional collaboration for NDC for scaling up resilience at a regional level. It further provides recommendations which include the alignment and integration of NDCs with other related processes, such as national adaptation plans (NAPs), SDGs, and the SFDRR.

3.3.2 Identifying and Addressing Components of L&D

Several chapters in this publication focus on better understanding the concept of climate-induced L&D. This includes identifying economic and non-economic losses and damages; slow-onset events; better understanding climate-induced migration in the region; as well as forms of addressing the losses and damages faced.

The chapter "Identification of non-economic loss and damage (NELD) indicators and practices in the context of climatic events," (Prabhakar et al., 2023a) focuses on enhancing the understanding of non-economic L&D and relevant indicators. It studies the understanding of different stakeholders including community and officials, and approaches for localizing the NELD indicators. The chapter looks at also community and participatory approaches related to L&D and the conceptual understanding of the thematic area.

The chapter "Priorities for addressing Slow-Onset Events (SOEs) in selected ASEAN countries," (Prabhakar et al., 2023b) focuses on another form of climate-induced L&D. The chapter focuses on slow-onset events and their impacts on the ASEAN member states that face significant implications on the countries' social and economic empowerment. The chapter also looks at the national priorities with a country focus on Cambodia, Malaysia, the Philippines, and Vietnam.

The chapter "Climate change-induced migration in South Asia" (Ranjan, 2023) looks at climate-induced migration which could be considered as impacted by climate change and disaster-related impacts. The South Asian region remains one of the key areas where migration remains high, and the chapter studies the implications of migration as well as economic and non-economic L&D related to migration. The chapter highlights the elevated need for addressing climate-induced migration in the region through policy and action.

3.3.3 Assessing and Addressing Climate Risk Through Community-Centric Approaches

Several chapters in the publication focus on assessing impact, with a focus on community-driven and centric, as well as participatory and multi-stakeholder-engaged approaches. Amongst these chapters are "A people-centred approach to assess and address impacts of climate change induced loss & damage," (Singh and Sharma, 2023) which elaborates on the Handbook on community-led assessment of climate-induced L&D which consists of participatory tools for communities to assess their losses and damages from climate change.

Also, the chapter "Pillars of resilience: Assessing loss and damage at the local level," by (Lotia, 2023) also discusses community-driven good practices such as the toolbox for the local-level assessment of L&D resulting from climate-related stressors, developed through a regional project under CAF. The chapter focuses on the impacts on multiple sectors and explores the opportunities for integrated approaches which could generate higher impact.

Another chapter under this thematic focus is "A conceptual framework and research design for assessing losses and damages from climate change in vulnerable communities," (Van der Geest, 2023) which is focused on the same conceptual framework as the previous chapter. It builds its insights from case studies and engagement with communities, and discusses different aspects related to climate-induced L&D, including but not limited to sudden and slow-onset events as well as coping strategies to the impacts faced.

The chapter "Climate change risk assessment and adaptation for loss and damage of urban transportation infrastructure in Southeast Asia," (by Noi et al., 2023) focuses on the assessments at a sectoral level, with a focus on the transportation infrastructure. It describes the assessment of adaptation processes in the sector and consists of different assessment components which also include L&D assessments that are focused on floods in three cities in Cambodia, Thailand, and Vietnam.

"Towards effective loss and damage systems in disaster recovery contexts in Southeast Asia," by (Boyland et al., 2023) is also a chapter that relates to this thematic focus. It examines the impacts of climate change and disasters in Southeast Asia. Expanding further on the assessment component, this chapter explores disaster recovery processes and L&D systems. The scope of the paper expands to the Vietnamese Mekong Delta; Prey Veng, Cambodia; Aceh, Indonesia; and Central Thailand. The chapter further provides intervention strategies aimed at synergized action related to disaster recovery and climate-induced L&D.

3.3.4 Forward-Looking and Innovative Approaches

Building on the evidence generated from fourteen CAF projects of the APN, the publication moves further in the present analysis to forward-thinking approaches for addressing the needs related to integration and synergizing related policy and action processes. The chapter titled "Forward-thinking perspectives towards integrating climate change adaptation and disaster risk reduction for synergetic resilience-building and addressing loss and damage," (Mombauer and Wijenayake, 2023) studies the different policy and strategic processes which provide entry points for scaled-up synergies for creating collective impact. It looks at both global and regional levels as well as innovative national and local-level approaches, as well as avenues for enhancing synergies and developing a holistic framework that interlinks CCA and DRR to effectively build resilience and address L&D.

With a multi-sectoral and multifaceted approach, the publication through contribution from insights generated from projects implemented across the Asia-Pacific region aims to provide informed approaches for creating synergies through the integration of processes to generate the maximum impact for communities vulnerable to the impacts of disasters and climate change. The publication is a collective effort of multiple actors with diverse technical expertise and representation of different stakeholder groups enabling the inclusion of diverse voices and views related to the overarching thematic focuses.

It is the hope and aim of editors and authors that through this effort, the readers may succeed in delving into different approaches for synergy-building; interlinked and strategic approaches for identifying key moments and opportunities for scaling up climate science-driven solutions to addressing climate risks, achieving SDGs, and building long-term resilience that are inclusive, participatory, and science-driven.

References

Basnayake, Senaka, Guota, Niladri, Rahman, Md. Habibur, Jayasinghe, Susantha, Ulubasoglu, Mehmet Ali, & Tong, Lan Anh (2023). *Flood-induced losses and damages to agricultural production and farmers' mental wellbeing in Sri Lanka*.

Boyland, Michael, Marks, Danny, Kimkong, Ham, & Nugroho, Agus (2023). *Towards effective loss and damage systems in disaster recovery contexts in Southeast Asia*.

Hoang, Hop T. B., Lasco, Rodel, Nguyen, Ninh H., & Boer, Rizaldi (2023). *SWOC analysis on the proposed linkage between Climate Change Adaptation (CCA), Disaster Risk Reduction (DRR), and Loss and Damage (L&D): Case Studies in the low-lying coastal cities of Indonesia, Philippines, Thailand and Vietnam*.

Intergovernmental Panel on Climate Change (2022). *Climate change 2022: Impacts, adaptation and vulnerability*. Working Group II contribution to the Sixth Assessment Report of the Intergovernmental Panel on Climate Change.

Lotia, Hina (2023). *Pillars of resilience: Assessing loss and damage at the local level*.

Mombauer, Dennis & Wijenayake, Vositha (2023). *Forward-Thinking perspectives towards integrating climate change adaptation and disaster risk reduction for synergetic resilience-building and addressing loss and damage*.

Nhat, Le Minh & Thinh, Dang Quang (2023). *Linkages between disaster risk reduction and climate change adaptation in the context of increasing climate change-induced loss and damage in Vietnam*.

Prabhakar, S. V. R. K., Chiba, Yohei, & Islam, Md. Atikul (2023a). *Identification of non-economic loss and damage (NELD) indicators and practices in the context of climatic events*.

Prabhakar, S. V. R. K., Pereira, Joy Jacequline, Pulhin, Juan M., Thang, Nguyen Van, Nyda, Chhinh, Aquino, Dannica Rose G., Nga, Pham Thi Thanh, Mau, Nguyen Dang, Thinh, Dang Quang, & Thuy, Tran Thanh (2023b). *Priorities for addressing Slow-Onset Events (SOEs) in selected ASEAN countries*.

Ranjan, Amit (2023). *Climate change-induced migration in South Asia*.

Singh, Harjeet & Sharma, Sindra (2023). *A people-centred approach to assess and address impacts of climate change induced loss & damage*.

Van der Geest, Kees (2023). *A conceptual framework and research design for assessing losses and damages from climate change in vulnerable communities*.

CHAPTER 2

Flood-Induced Losses and Damages to Agricultural Production and Farmers' Mental Wellbeing in Sri Lanka

Senaka Basnayake, Niladri Gupta, Muhammad Habibur Rahman, Susantha Jayasinghe, Mehmet Ali Ulubasoglu, and Lan Anh Tong

Abstract This study focuses on the impact of floods on agricultural yields and farmers' mental wellbeing in flood-prone areas of Sri Lanka.

S. Basnayake (✉) · N. Gupta · S. Jayasinghe
Asian Disaster Preparedness Center, Bangkok, Thailand
e-mail: senaka_basnayake@adpc.net

N. Gupta
e-mail: niladri.gupta@adpc.net

S. Jayasinghe
e-mail: susantha@adpc.net

M. H. Rahman
Department of Economics, Durham University Business School, Durham University, Durham, UK
e-mail: habib.rahman@durham.ac.uk

V. Wijenayake et al. (eds.), *Linking Climate Change Adaptation, Disaster Risk Reduction, and Loss & Damage*,
https://doi.org/10.1007/978-981-99-8055-0_2

Food security is facing heightened risks with the increasing frequency and severity of natural hazards globally, including storms, floods, droughts, extreme temperatures, heat waves, and sea-level rise. To assess these risks, the study conducted a data collection exercise in Sri Lanka, gathering detailed information on crop production, farmers' demographics, and agricultural inputs. The collected data will be used for future economic analyses to evaluate potential losses and damages caused by floods and to understand the likely effects on the wellbeing of affected farmers. The study aims to provide insights into climate-inclusive agricultural risk assessment and resilience mechanisms for farmers facing flood-related challenges. By understanding the potential impact of floods on agricultural production and farmers' psychological wellbeing, strategies can be developed to mitigate risks and enhance resilience in flood-prone areas.

Keywords Losses and damages · Wellbeing · Floods · Agriculture · Sri Lanka · Disaster resilience

Highlights

- Farmers in flood-struck and flood-free areas in Sri Lanka have the same demographic characteristics.
- However, they differ in their agricultural production and likely in their psychological wellbeing.
- It is likely that farmers in flood-struck areas employ several coping mechanisms to floods with varying impacts based on their demographic characteristics.

M. A. Ulubasoglu · L. A. Tong
Deakin University, Burwood, VIC, Australia
e-mail: mehmet.ulubasoglu@deakin.edu.au

L. A. Tong
e-mail: lan.tong@deakin.edu.au

1 Introduction

Climate change is inevitable. Agriculture, which determines the food security of billions of people globally, is one of the most vulnerable sectors to climate change. Agriculture is an open system and provides food and employment for 60% of the world's population, with more than 2.2 billion people in Asia depending on it for their livelihoods.

Due to the distinct climatic variability across the continent and its geophysical setting, most countries in Asia are subject to natural disasters. The frequency of these extreme events, especially that of hydrometeorological events (e.g., floods, droughts), has shown an increasing trend (Olano 2022). Recent studies have shown that the world is moving towards scenarios of either too much, too little, or erratic water availability due to climate change. Thus, climate-induced natural hazards are likely to considerably affect the agricultural sector as well as the livelihoods of the population depending on it.

The present study aims to shed light on the consequences of flood events on farming communities in Sri Lanka to understand potential impact of floods on agricultural yields in Sri Lanka as well as to discuss potential ways to understand the effects of floods on psychological wellbeing of Sri Lankan farmers. We collect data from different parts of the country to provide preliminary information about the ways in which farmers undertake agricultural production in flood-prone vis-a-vis flood-free agricultural zones. Our objective is to enable a climate-inclusive agricultural risk assessment in flood-prone areas with a view to undertaking economic analyses to evaluate the potential losses and damages as well as to discuss the likely impact on mental wellbeing of affected farmers, suggesting resilience mechanisms for the likely effect of floods on farmers.

1.1 Climate Risk and Disaster Management Vis-a-Vis Loss and Damage

One of the major shortfalls of conventional disaster management strategies is a lack of adequate blends of climatic information on the nature of future climate risks and post-disaster reconstruction processes or modalities. This lack of information is more acute in agricultural communities in Asia. The agricultural sector is adversely affected by disasters in a multitude of ways, with lower-than-expected production being the primary impact. A cascading effect also occurs starting from direct economic loss to farmers along the entire value chain, even affecting the growth of the sector or entire national economies (FAO 2021).

Losses and damages (L&D), on the other hand, are often divided into economic losses and damages, including livelihoods and property, and non-economic losses and damages (NELD), which can comprise loss of life, negative impacts on mental health, and the loss of biodiversity and cultural heritage. The estimation of NELD has emerged as a new concept in the negotiations under the United Nations Framework Convention on Climate Change, given that this L&D often goes unnoticed by the external world. As such, quantifying the economic L&D due to disasters is crucial for evidence-based decision-making to develop and improve the resilience mechanisms to extreme events. Sri Lanka is identified as an important area in the context of the discussions around the Nationally Determined Contributions (NDCs) of the United Nations Framework Convention on Climate Change (UNFCCC). This framework stresses the need for improving forecasting abilities, assessing L&D caused by climate-induced disasters, establishing local mechanisms in line with the Warsaw International Mechanism (WIM), and strengthening the existing national institutions to recover L&D, which includes introducing possible insurance schemes. From a disaster management perspective, post-disaster evaluation is carried out for estimating L&D for reconstruction and rehabilitation, while from a climate change adaptation perspective, evaluation or assessment is carried out to estimate L&D to minimize impacts of future events. In both cases, evaluation for the agriculture sector is very challenging (Basnayake et al. 2021). The emergence of NELD in the UNFCCC context presents an excellent opportunity to increase awareness of a frequently overlooked aspect of climate impacts. It is necessary to orchestrate the integration of research findings from various disciplines to create a robust, evidence-based foundation for devising fair and effective approaches to tackle NELD (Olivia et al. 2016).

1.2 The Psychological Impact of Floods

Several disciplines have been studying climate change and climate variability and the impact of resulting extreme events on natural resources. Extreme rainfall and associated floods caused by climate change have had significant impacts worldwide (Blöschl et al. 2017; Ciscar et al. 2011), not only damaging structures in sectors like agriculture, infrastructure, and natural resources, but also affecting people's psychological wellbeing (Fothergill et al. 2021). As a result, there has been a rise in the number of studies that use individual-level happiness or life satisfaction data to measure the effects of extreme weather events on happiness (Di Tella and MacCulloch 2006; Frey and Stutzer 2012).

In 2009, Luechinger and Raschky conducted one of the initial studies in this field, in which they used life satisfaction data to estimate and evaluate utility losses due to floods in 16 European countries from 1973 to 1998. The study revealed a substantial negative effect of floods on life satisfaction and suggested that risk transfer methods, such as mandatory insurance, could significantly alleviate the consequences. The emergence of NELD in the recent past has increased the necessity to include NELD in studies on L&D.

This study shares some initial thoughts on the potential methodologies (e.g., panel-based econometric studies) for estimating L&D due to floods in the agricultural sector as well as the need to estimate the impact of flooding on subjective wellbeing of Sri Lankan farmers.

1.3 Study Context

Sri Lanka is an island located at the tip of the Indian sub-continent surrounded by the Indian Ocean. Of Sri Lanka's rural population, 77.4% depends on agriculture as their major livelihood activity (Marambe et al. 2017), and these communities are frequently hit by natural disasters. Sri Lanka has seen a steady rise in the frequency of floods over the past two decades (IWMI 2018). Flood occurrence in Sri Lanka was almost regular in the recent decade (every year over the last ten years), with an apparent increase in the affected areas. While, on average, fewer than five districts were affected by floods during the period 1991–2000, the last decade has seen a doubling of the geographic coverage of floods. Against the backdrop of the limited number of studies on the effect of floods on agricultural production in Sri Lanka, understanding the effect of floods on agricultural production and how farmers respond to the L&D caused by floods is vital for policymakers and relevant government institutions for developing better coping mechanisms to disasters. These studies can be further extended to understand causal effects of floods on life satisfaction of the farmers. It is understood that farmers generally experience major monetary and psychological impacts due to floods because their livelihood depends on soil, livestock, and agricultural crops. It is known that floods can lead to significant damages on soil quality and losses in agricultural crops and livestock. Sometimes, such losses may not be restricted to a couple of months after the floods, but the negative consequences of floods can be experienced for several years. Some proposed way forward to estimate these NELD has been discussed which is a novel contribution to this article.

2 Research Design

2.1 Treatment and Control Groups

A standard randomized sampling technique developed by Deakin University in Australia was employed to select the pilot sites, including flood-affected areas (treatment sites) and non-flood-affected areas (control sites). Sri Lanka has a total of 25 districts, which can be categorized into three zones: wet, intermediate, and dry. In the first stage of the research design, based on the previous 25 years (1991–2015) of flood history, districts are divided into a more-affected group, which comprises those with more than four years of flood incidents, and a less-affected group for the districts with less than four years of flood experience. Based on this categorization, 11 districts fall into the more-affected group, leaving 14 remaining districts in the less-affected group. We further eliminate 8 districts in the dry zone as the dry zone contains well-established irrigation systems making these districts not comparable to the districts in the wet and intermediate zone. So, for the more-affected group, this leaves 11 districts with ten districts belonging to the wet zone and one district belonging to the intermediate zone, while the less-affected group includes 6 districts belonging to the wet and intermediate zones. This selection procedure guarantees that all farmers included in our treatment and control groups use similar irrigation systems, which is a crucial factor for agricultural practice. The research design randomly chose four districts, two potential treatment districts from the more-affected group (Galle and Gampaha) and two control districts (Kegalle and Monaragala) from the less-affected group (see Fig. 1 for the location of the treated and control groups). In the second stage, the land use map from the Survey Department of Sri Lanka is utilized to identify all agricultural plots. From each of the four chosen districts, 125 agricultural plots are randomly selected. The spatial distribution of the selected plots in the fragmented land used for agriculture is shown in Fig. 1. The cultivators or household heads of each plot are then identified and surveyed with the details being presented in the next section.

Our procedure and criteria in selecting appropriate treatment and control groups are below:

1. Collecting district-level historical flood data (affected area and the timeline of flood events, number of casualties, and assessed L&D);
2. Dividing all 25 districts into two groups: experiencing >10 floods (potentially treated districts) and exposed to at most about 3–4 floods (potentially control districts) in the last 25 years;

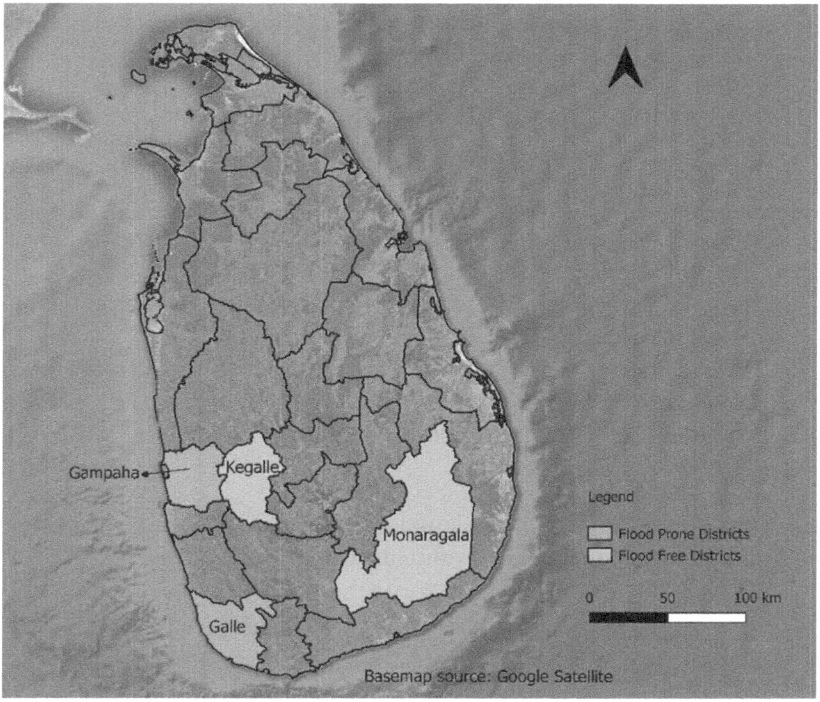

Fig. 1 Treatment districts and control districts for the study

3. Collecting the land use/landcover data in GIS format (shapefiles) that include information such as arable/farmland and the type of crops cultivated in each unit of land;
4. Overlaying district administrative boundaries on the land use map;
5. Within the list of potentially treated districts (in point 2), randomly select two districts, which happened to be Galle and Gampaha;
6. For each of the two districts selected as per criterion 5, randomly select agricultural units and ensure that: (a) The number of plots is approximately equal across two selected districts and (b) those plots' total area should sum up to a similar figure even though there could be variations;
7. Information was also collected on the characteristics for each of the randomly selected rice plots in criterion 6 on the size of cultivated land, proximity to water body, forest land, elevation/ruggedness, rainfall and temperature statistics, socio-economic status of the households, and on the whole, paddy-related data.

Another two districts were also randomly selected as control districts, which happened to be Kegalle and Monaragala. The selection of agricultural units meets the criteria as mentioned below:

a. The selection of the rice plots should be such that they resemble approximately the size of the treatment plots;
b. The number of selected rice plots for treatment and control is kept equal and considers the geographic, topographic, demographic, economic, and socio-economic attributes the same for all the districts.

Based on the above, 125 rice plots were considered from each district randomly in both control and treatment groups.

2.2 Conducting the Baseline and Endline Surveys

Our baseline survey was conducted at the beginning of the paddy season to avoid any potential effects of floods on the cultivation process such as field preparation, transplantation, and maintenance. In Sri Lanka, paddy is grown in two seasons, Maha (main season with cultivation time from September to March) and Yala (secondary season with cultivation time

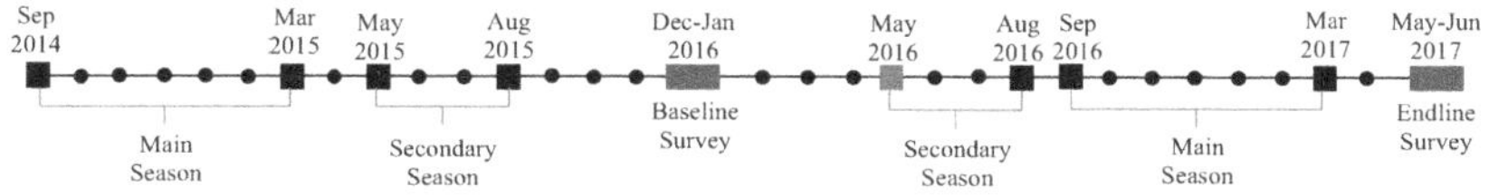

Fig. 2 Temporal schematic of the survey

from May to August). As our chosen districts are in the Southwest region that is prone to monsoon rains and occasional floods during May–September, we conducted the baseline survey in December 2015–January 2016, prior to the Yala season. The survey covered 500 respondents, with 125 cultivators selected from each of the four districts. After the baseline survey, we predicted that Galle and Gampaha would likely be affected by floods during the Yala season, while Kegalle and Monaragala would not. This prediction turned out to be accurate, as Galle and Gampaha were flooded in May 2016, while Kegalle and Monaragala were mostly unaffected. The endline survey was conducted in May–June 2017, one year after the flooding (see Fig. 2 for a temporal schematic of the surveys), with 100 cultivators from Galle (25 missing cultivators), 124 from Gampaha (1 missing cultivator), and 125 from each of the control districts. Of the 125 cultivators in each district, 65 in Galle and 61 in Gampaha were actually affected by the 2016 flood. Thus, our sample size becomes 948 observations of which 252 observations are included in the treatment group and the remaining 696 observations in the control group.

2.2.1 Farmer Characteristics: Overall Sample

Table 1 displays descriptive statistics for the overall sample which extracts from the survey of 948 household respondents in years before (2015) and after the flood (2017). The characteristics are classified into demographic characteristics of households, demographic characteristics of household heads, economic characteristics of households, and agriculture-related characteristics. For the first category, on average a household in the sample has 4.4 persons, but the number of household members varies from 2 to 12. The number of males in a household is 2.2, which is almost 50% of total household members (4.4 persons as shown in cell [1] of Table 1). The number of household members aged above 50 is 1.74, accounting for almost 40% of household members. Turning to the second category, the demographic characteristics of household heads, 92% of households had male household heads, which is consistent with male dominance in family in developing countries and rural areas. Average age of household head is quite high, at 56.5 years on average, but also varies from as young as 25 up to 93 years. The proportion of married and unmarried household head is similar (51% as opposed to 48%, with 1% being widow). All household heads are literate and mostly (90%) hold primary (45%) or secondary school (45%) level as highest level of education leaving 10% with tertiary education. Overall, the demographic characteristics show that a household in the sample has a medium size, and household members including household heads are old. In addition, household heads are mostly male, aged above 50, and have primary or secondary level of education.

Regarding economic characteristics, a household owns on average around 8,387 metre^2 of total land (both agricultural and non-agricultural land) and earns on average \$49.5 per week. For agriculture-related outcomes, almost all households in the sample cultivate paddy, while only 35% of households cultivate non-paddy crops along with paddy (85 households out of 948 surveyed households do not report paddy production outcomes). Non-paddy crops include more than 30 varieties, but most common are corn, tea, banana, pepper, and vegetables. This is common in Asian agriculture with small household farming. Rice is the main source of calories and the primary crop for subsistence owing to its less variability in crop yield and being the direct source of food for households. Cash crops, on the other hand, can be more profitable but also more volatile in yield and market price. In our sample, on average, each household cultivates 6378 meter^2 of paddy per year with non-paddy production area being slightly less given that cash crops such as banana, coconut, tea often need larger space. Average paddy yield is 0.5 kg per metre^2, but the yield varies from 0.01 to 3.12 kg per metre^2 across households and years while average yield for non-paddy crops is 1.47 kg per metre^2. Cultivation duration for crops is 7 months while fertilizer use for all crops is 670 kg. The number of crops produced is 1.42 which is consistent with the fact that most households produce just rice and around 35% of them produce other crops along with rice.

2.2.2 *Characteristics of Treatment and Control Groups*

It can be seen from Table 2 that the demographic characteristics of households and those of household heads are similar for the treatment and control groups. This is illustrated in column (9) in that the difference in the mean values of these characteristics is very small. The only observable

Table 1 Descriptive statistics (overall sample)

	Mean	*Standard derivation*	*Min*	*Max*	*Count*
Demographic Characteristics of Households:					
Household Size	4.39	1.37	2.00	12.00	919
Number of Household Male	2.20	0.94	1.00	6.00	935
Household Members Aged 50+	1.74	0.58	1.00	6.00	698
Demographic Characteristics of Household Heads:					
Age of Household Head	56.48	12.75	25.00	93.00	943
Male	0.92	0.27	0.00	1.00	948
Married	0.51	0.50	0.00	1.00	948
Unmarried	0.48	0.50	0.00	1.00	948
Literate	1.00	0.03	0.00	1.00	942
Primary level	0.45	0.50	0.00	1.00	936
Secondary level	0.45	0.50	0.00	1.00	936
Tertiary level	0.10	0.29	0.00	1.00	936
Economic Characteristics:					
Land Area Owned	8387.77	9451.61	180.18	120,000.00	900
Weekly Income	49.51	29.97	7.14	297.80	907
Agriculture-related Characteristics:					
Paddy Production	2693.01	2813.45	29.92	23,000.00	863
Paddy Production Area	6378.59	6516.87	180.18	72,843.00	863
Paddy Yield	0.50	0.33	0.01	3.12	863
Non-paddy Production	4703.82	9936.08	4.00	105,000.00	302
Non-paddy Production Area	5831.21	8532.89	25.29	96,000.00	302
Non-paddy Yield	1.47	3.39	0.00	28.00	302
Cultivation Duration	7.08	5.72	1.00	36.00	869

(continued)

Table 1 (continued)

	Mean	*Standard derivation*	*Min*	*Max*	*Count*
Volume of Fertilizer	670.04	1191.06	10.00	15,000.00	869
Number of Crops Produced	1.42	0.61	1.00	3.00	877

Notes This table provides the average characteristics of the whole sample, including responses from 948 households. Household size is the number of household members; literacy is male, married, or unmarried; primary and secondary variables are indicators equal to 1 if the household head is literate; being male, being married, or unmarried; having primary or secondary school as the highest level of education; and zero otherwise; land area owned is the total land owned by the household in $metre^2$ while weekly income is in US dollars. Cultivation period is in months and volume of fertilizer is in kg. Crop output is measured in kg; crop area is in $metre^2$; crop yield is kg/m^2

difference is the total land area, which is about 7.2% higher in the treatment group than the control group. This attests to our research design as a randomized control trial resulting in a balanced sample between treated and non-treated groups in terms of their characteristics not directly related to the agriculture production.

In contrast to the above-mentioned characteristics, there are some noticeable differences between the treatment and control groups in agriculture-related characteristics. Paddy production area cultivated by a household in flood-struck areas is 13.3% higher than that in non-flood-affected areas. The paddy yield of the former is just above half of the latter, resulting in substantial lower output of almost 19%. On the other hand, production area for non-paddy crops is 24% lower in flood-struck area compared with the control group. Nonetheless, there is no difference in yield of non-paddy crops. The results suggest that the flood-struck areas are different in their paddy production, but not in non-paddy production. For other outcomes, the difference is observed for fertilizer use, with households in flooded-struck applying almost 10% higher fertilizer than those in non-flooded areas. The higher fertilizer use can be a way farmer uses to cope with productivity loss caused by the floods. In short, the raw data show significantly lower productivity and output loss for paddy production for the affected households, and farmers tend to increase the production area and use more fertilizer to possibly reverse the productivity loss.

3 Discussion

The present study is significant as well as unique in using a research design that facilitates more accurate and timely assessment of L&D caused by natural hazards. The use of climate models to carry out climate-inclusive hazard assessment in the future can be used effectively to avert, minimize, and address L&D from climate change, which is a continuous process. It is also important that under all assessed scenarios for mitigation and adaptation, damage from residual risk is unavoidable (Mechler et al. 2020). Additionally, L&D refers to impacts of climate-related stressors that have not been or cannot be avoided through mitigation and adaptation. Thus, planning becomes essential especially for the decision-makers in the agriculture sector.

The above data acquired on treatment and control groups can also be used as an input to panel-based econometric models (Ashenfelter and

Table 2 Descriptive statistics (by treatment status)

	Treatment group				*Control group*				*Difference*
			(1)-(5)						
	(1) *Mean*	*(2)* *Standard derivation*	*(3)* *Min*	*(4)* *Max*	*(5)* *Mean*	*(6)* *Standard derivation*	*(7)* *Min*	*(8)* *Max*	*(9)*
Demographic Characteristics of Households:									
Household Size	4.30	1.27	2.00	9.00	4.43	1.40	2.00	12.00	−0.13
Number of Household Male Males	2.08	0.89	1.00	5.00	2.25	0.95	1.00	6.00	−0.17
Household Members Aged 50 +	1.74	0.51	1.00	4.00	1.74	0.60	1.00	6.00	0
Demographic Characteristics of Household Head:									
Age of Household Head	55.64	12.31	25.00	80.00	56.78	12.90	25.00	93.00	−1.14
Male	0.96	0.20	0.00	1.00	0.91	0.28	0.00	1.00	0.05
Married	0.49	0.50	0.00	1.00	0.52	0.50	0.00	1.00	−0.03
Unmarried	0.51	0.50	0.00	1.00	0.47	0.50	0.00	1.00	0.04
Literate	1.00	0.00	1.00	1.00	1.00	0.04	0.00	1.00	0
Primary level	0.37	0.48	0.00	1.00	0.48	0.50	0.00	1.00	−0.11
Secondary level	0.57	0.50	0.00	1.00	0.41	0.49	0.00	1.00	0.16
Tertiary level	0.06	0.23	0.00	1.00	0.11	0.31	0.00	1.00	−0.05
Economic Characteristics:									
Land Area Owned	8818.52	7187.14	505.50	40,467.00	8226.65	10,170.12	180.18	120,000.00	591.87

	Treatment group				*Control group*				*Difference*
	(1) Mean	*(2) Standard derivation*	*(1)-(5) (3) Min*	*(4) Max*	*(5) Mean*	*(6) Standard derivation*	*(7) Min*	*(8) Max*	*(9)*
Weekly Income	49.64	23.37	7.14	152.00	49.46	32.14	7.50	297.80	0.18
Agriculture-related Characteristics:									
Paddy Production	2367.67	2532.85	290.00	19,200.00	2810.53	2901.08	29.92	23,000.00	−442.86
Paddy Production Area	6981.97	6443.80	1000.00	48,560.00	6160.65	6534.41	180.18	72,843.00	821.32
Paddy Yield	0.35	0.18	0.07	1.61	0.55	0.36	0.01	3.12	−0.2
Non-paddy Production	5056.55	9706.04	0.00	75,200.00	4280.60	9814.04	0.00	105,000.00	775.95
Non-paddy Production Area	4878.54	8028.50	100.00	61,714.00	6046.44	8657.92	0.00	96,000.00	−1167.9
Non-paddy Yield	1.44	2.81	0.00	20.00	1.44	3.58	0.00	28.00	0
Cultivation Duration	7.11	6.95	1.00	36.00	7.07	5.21	1.00	36.00	0.04

(continued)

Table 2 (continued)

	Treatment group				*Control group*				*Difference*
			(1)-(5)						
	(1)	*(2)*	*(3)*	*(4)*	*(5)*	*(6)*	*(7)*	*(8)*	*(9)*
	Mean	*Standard derivation*	*Min*	*Max*	*Mean*	*Standard derivation*	*Min*	*Max*	
Volume of Fertilizer	716.63	1271.67	15.00	10,000.00	652.98	1160.71	10.00	15,000.00	63.65
Number of Crops Produced	1.46	0.63	1.00	3.00	1.41	0.60	1.00	3.00	0.05

Notes This table provides the average characteristics and the difference in the mean value of the treatment and control group farmers. Household size is the number of household members; literacy is male, married, or unmarried; primary and secondary variables are indicators equal to 1 if the household head is literate; being male, being married, or unmarried; having primary or secondary school as the highest level of education; and zero otherwise; land area owned is the total land owned by the household in metre^2 while weekly income is in US dollars. The cultivation period is in months and volume of fertilizer is in kg. Crop output is measured in kg; crop area is in metre^2; crop yield is kg/m^2

Card 1985; Bertrand et al. 2004; Wing et al. 2018) to quantify the effect of floods on agricultural production (Merz et al. 2010) and understand behavioural characteristics of farmers who are averse to risk-taking and farmers who show a willingness to take risk.

Similarly, the causal effect of natural disasters, specifically floods, on subjective wellbeing can also be modelled by similar panel-based approach by exploiting the plausibly exogenous variation in living in a flood-exposed district and potential exposure of flood during the period between the baseline and endline surveys. The control for demographic characteristics (gender, age, education, marital status) can be considered in the model, as well as the control for potential endogenous variables (risk-taking, agricultural output, income, and wealth indicators) to estimate the flood effects on life satisfaction.

Agricultural production largely depends on weather conditions and is extremely prone to natural hazards. A more frequent and severe occurrence of natural hazards such as storms and floods in recent decades have put food security at an increased risk, especially for farmers in least developed countries who have access to limited resources to cope with abrupt and adverse events. Sri Lanka, a developing country with much of the rural population being dependent on agriculture, is hit by annual floods with a recent increase in affected areas. This study collects data to shed light on the effect of floods on agricultural production and farmers' response to flood incidents. Our research design is based on the principles of randomized control trial to construct a treatment and control group at district level. We classify flood-prone and flood-free district groups based on the frequency of their exposure to floods in the past 25 years and their similarity in agricultural facility to ensure that the treatment and control groups offer comparable information other than their flood exposure.

Our findings show that our treatment and control groups are mostly similar in characteristics that are not directly related to agricultural production, attesting that our classification is appropriate. The data show that households in flood-hit areas (treatment group) experience drastically lower paddy productivity (57%) compared with those in non-flood-hit areas. In addition, they tend to expand the land area for paddy production and use more fertilizer, potentially to make up for their output loss. The above methodology can be used to estimate potential L&D to agricultural production that may occur in the future using more robust econometric models. The models can further include parameters to understand the causal effect of natural disasters, specifically floods, on subjective wellbeing. The methodology, if incorporated in a Decision Support System (DSS), would enable future decision-making as well as better management of impact on farmers' livelihood as well as farmer's psychological wellbeing due to floods as well as other hydroclimatic hazards like drought. The study also has implications for policy formulation on managing risks to the agriculture sector as well as to provide support to affected households.

Acknowledgements The research was funded by the Asia-Pacific Network for Global Change Research (CAF2014-RR01NMY-Basnayake). The authors would like to thank all the participants for providing their time and knowledge during the survey carried out in Sri Lanka. The author would also like to express gratitude to all project partners and their affiliated organizations for the support in carrying out the research.

References

Ashenfelter, Orley, and Card, David. 1985. "Using the Longitudinal Structure of Earnings to Estimate the Effect of Training Programs." The Review of Economics and Statistics, 67(4), 648. https://doi.org/10.2307/1924810.

Basnayake, Senaka, Ulubasoglu, Mehmet, Rahman, Muhammad Habibur, Premalal, Sarath, Chandrapala, Lalith, Shrestha, Madan Lall, Jayasinghe, Susantha, and Gupta, Niladri. 2021. "Assessing potential loss and damage for flood hazard using an econometric modelling technique." APN Science Bulletin, 11(1). https://doi.org/10.30852/sb.2021.1499.

Bertrand, M., Duflo, E., and Mullainathan, S. 2004. "How much should we trust differences-in-differences estimates?" The Quarterly Journal of Economics, 119(1), 249–275. https://doi.org/10.1162/003355304772839588.

Blöschl, G., Hall, J., Parajka, J., Perdigão, R.A., Merz, B., Arheimer, B., Aronica, G.T., Bilibashi, A., Bonacci, O., Borga, M. and Čanjevac, I. 2017. Changing climate shifts timing of European floods. Science, 357(6351), 588–590.

Ciscar, J.C., Iglesias, A., Feyen, L., Szabó, L., Van Regemorter, D., Amelung, B., Nicholls, R., Watkiss, P., Christensen, O.B., Dankers, R. and Garrote, L. 2011. Physical and economic consequences of climate change in Europe. Proceedings of the National Academy of Sciences, 108(7), 2678–2683.

Di Tella, R. and MacCulloch, R. 2006. Some uses of happiness data in economics. Journal of Economic Perspectives, 20(1), 25–46.

FAO. 2021. The impact of disasters and crises on agriculture and food security: 2021. Rome. https://doi.org/10.4060/cb3673en.

Frey, B.S. and Stutzer, A. 2012. The use of happiness research for public policy. Social Choice and Welfare, 38(4), 659–674.
Fothergill, L.J., Disney, A.S., and Wilson, E.E. 2021. A qualitative exploration of the psychological impacts of living with the uncertainty of persistent flood risk. Public Health, 198, 141–145. https://doi.org/10.1016/j.puhe.2021.07.016. Sep. Epub 2021 Aug 20. PMID: 34425284.
IWMI. 2018. Getting ahead of disaster risks. Retrieved from https://www.iwmi.cgiar.org/2018/08/gettingahead-of-disaster-risks/.
Marambe, Buddhi, Silva, Pradeepa, and Athauda, Saman. 2017. "Agriculture and rural development under central government and provincial council setup in Sri Lanka." Decentralization and Development of Sri Lanka Within a Unitary State, 111–145. https://doi.org/10.1007/978-981-10-4259-1_6.
Mechler, R., Singh, C., Ebi, K., Dialante, R., Thomas, A., James, R., and Tschakert, P. 2020. Loss and damage and limits to adaptation: recent IPCC insights and implications for climate science and policy. Sustainability Science, 15(4), 1245–1251.
Merz, B., Kreibich, Heidi., Schwarze, Reimund., and Thieken, Annegret. (2010). Review article "Assessment of economic flood damage". *Natural Hazards and Earth System Sciences 10*. https://doi.org/10.5194/nhess-10-169-2010.
Olano, G. 2022. Managing the rising tide of flood risk in Asia. Insurance Business Asia. Source: https://www.insurancebusinessmag.com/asia/news/breaking-news/managing-the-rising-tide-of-flood-risk-in-asia-414039.aspx.
Olivia, S., Waters, E., and Chan, S. 2016. Non-economic loss and damage: addressing the forgotten side of climate change impact. Briefing Paper 3, German Development Institute: Bonn. Source: https://www.idos-research.de/en/briefing-paper/article/non-economic-loss-and-damage-addressing-the-forgotten-side-of-climate-change-impacts/.
Wing, C., Simon, K., and Bello-Gomez, R.A. 2018. Designing difference in difference studies: best practices for public health policy research. Annual Review of Public Health, 39(1), 453–469. https://doi.org/10.1146/annurevpublhealth-040617-013507.

CHAPTER 3

A People-Centred Approach to Assess and Address Impacts of Climate Change-Induced Loss and Damage

Harjeet Singh and Sindra Sharma

Abstract For over a decade, the United Nations Framework Convention on Climate Change (UNFCCC) has been discussing the issue of Loss and Damage, yet little research has been done on frontline communities impacted by climate change. To address this, the Asia-Pacific Network for Global Change Research (APN) developed a Handbook on community-led assessment of climate-induced Loss and Damage, providing participatory tools for communities to assess their losses and damages from climate change. The Handbook is increasingly relevant as climate impacts increase and disaster risk reduction and adaptation efforts need to be scaled up. This paper examines how community-based

H. Singh · S. Sharma (✉)
Climate Action Network International, Bonn, Germany
e-mail: ssharma@climatenetwork.org

H. Singh
e-mail: hsingh@climatenetwork.org

V. Wijenayake et al. (eds.), *Linking Climate Change Adaptation, Disaster Risk Reduction, and Loss & Damage*,
https://doi.org/10.1007/978-981-99-8055-0_3

Loss and Damage assessments can benefit a subset of processes that have evolved under the UNFCCC and its Paris Agreement, including the Global Stocktake, Enhanced Transparency Framework, Nationally Determined Contributions (NDCs), and the Santiago Network for Loss and Damage.

Keywords Loss and Damage · Community-led assessments · Paris Agreement · Global Stocktake · Santiago Network · Nationally Determined Contributions (NDCs)

1 Introduction

Loss and Damage as discussed within the multilateral climate process is technical in nature and can be divorced from the reality of climate change impacts on the ground. Loss and Damage has been debated with incremental progress for the past three decades, led by the most vulnerable and obstructed by the developed countries. However, the Intergovernmental Panel on Climate Change (IPCC) Working II Report on climate change impacts (Pörtner et al. 2022) clearly shows the snowballing of impacts that the world is now facing, with the hard limits of adaptation being reached increasing the salience of unavoidable impacts. Practitioners have a vital role to play in bringing needs and evidence of lived experiences of affected communities to the forefront of the multilateral climate change decision-making space. Currently for the most part, practitioners and those impacted are largely excluded from the decision-making table with civic space worryingly shrinking in these vital venues (Santos et al. 2021). Civil society, however, has been engaging the levers they have to ensure that the voices of the impacted communities are heard inside the negotiation rooms.

For Loss and Damage at COP26, Civil Society delivered a substantive outcome of finance to address Loss and Damage, which was the litmus test of success (Singh 2021). At COP26, the negotiating bloc of G77 and China backed by the Alliance of Small Island States (AOSIS) proposed the delivery of a finance facility to address Loss and Damage. The compromise of the Glasgow Dialogue for Loss and Damage was instead reached. AOSIS made it clear that the aim of this Dialogue was to consultatively define the facility and make a recommendation to COP27. The first Dialogue took place during the Subsidiary Bodies meeting in Bonn in June 2022. G77 and China, AOSIS, and other developing nation blocs made it plain that progress was insufficient (AOSIS, 2022). The Dialogue provided no real direction for moving forward in terms of providing funding to remedy Loss and Damage Whilst the political outcome of COP26 and subsequently the first Glasgow Dialogue failed against this test, a critical and notable shift was seen in public awareness and opinion as well as political will with Scotland being the first country to pledge £1 million in support to victims of climate disasters (Lo and Farand 2021). And significant to this was the ability to forefront the experiences of those on the ground and to enable argumentation based on their needs.

With the weight of this momentum and the push of civil society, which kept forefronting the voices of the vulnerable and affected communities, there were two significant outcomes at COP27 in relation to Loss and Damage. One was on the technical arm of Loss and Damage—wherein it was decided to establish an Advisory Board of the Santiago Network as part of the Warsaw International Mechanism for Loss and Damage. The rules of procedure will only be adopted in 2024 which means the operationalisation of the Network still has some time to wait.

The second and arguably most significant outcome of COP27 was the long-fought issue of finance to address Loss and Damage. COP27 finally decided to establish new funding arrangements for assisting developing countries that are particularly vulnerable including with a focus on addressing Loss and Damage and that these new arrangements should complement and include sources, funds, processes, and initiatives under and outside UNFCCC and its Paris Agreement. It also decided to establish a fund for responding to Loss and Damage whose mandate "includes" a focus on addressing Loss and Damage. And lastly it decided to establish a Transitional Committee which would essentially iron out the details of the funding arrangements and make recommendations at COP28 for elements for operationalisation (UNFCCC 2022a). Now the key work begins to ensure that which is agreed is relevant to the people it is designed to serve.

The Handbook for Community-Led Assessment of Climate-Induced Loss and Damage (Anderson and Singh 2020) can be vital to these processes. The Handbook is a participatory tool that enables communities to assess the losses and damages they have experienced because of climate change and understand the trends and future changes that climate change impacts may bring. Such an assessment can help communities instigate strategic planning for future and current risks whilst also allowing them to systematically assess and record the economic and non-economic losses and damages they have experienced. This can be used to:

- Take steps to avoid or reduce future Loss and Damage by understanding trends related to climate change;
- Enable local and national authorities to understand and map the trends and impacts of climate disasters, and to plan to avoid future disasters through the provision of information;
- Engage with government to access relief, support, or compensation for recovery and rehabilitation in accordance with the assessment of climate-induced Loss and Damage;
- Compile evidence of climate-induced Loss and Damage so that the national government can demand support from the international community.

This research focuses specifically on the last—the critical role that the Handbook can play in shaping global climate policy that is needs-based and relevant to the people suffering impacts on the ground.

2 Methodology

2.1 Qualitative Research Methodology

To examine how the Loss and Damage Handbook is relevant to the United Nations Framework Convention on Climate Change (UNFCCC) policy space, the research involved a comparative analysis of the Handbook with key Paris Agreement processes. The research relied on emergent synthesis review a, qualitative research method that involves synthesising and integrating existing research on a particular topic or research question in a flexible and iterative manner. In this case, the research question was: How can procedures related to Loss and Damage under the UNFCCC and its Paris Agreement benefit from community-based Loss and Damage assessments? The methodology also utilised document analysis to gain a more comprehensive understanding of the research question. This involved analysing and interpreting written materials, specifically UNFCCC conference decisions from COP26 and COP27 and conference notes from COP27.

Specifically, document analysis was employed to identify and review a wide range of relevant documents, and then emergent synthesis review was used to synthesise and integrate the information in a flexible and iterative manner.

2.2 Selection of Processes Under the UNFCCC and the Paris Agreement

The scope of this research was to understand how community-based assessment of Loss and Damage is applied or could be applied to the processes of the UNFCCC and its Paris Agreement and specifically the applicability of the Handbook. Generally, needs-based Loss and Damage assessment can be seen as the process of identifying and assessing the needs of the developing countries that are facing Loss and Damage due to the impacts of climate change. This would include evaluating the impacts of climate change on these countries, such as increased frequency and severity of natural disasters, and determining the financial and other resources that will be needed to address these impacts. The community lens is integral to enable an accurate, context-bound assessment of needs.

One aspect of needs-based Loss and Damage assessment is the Paris Agreement's recognition of the need for the developed countries to provide financial and technical assistance to the developing countries to help them adapt to the impacts of climate change and to address Loss and Damage associated with those impacts. This is captured in several processes of the Paris Agreement which we identify here as the National Determined Contributions (NDCs), Global Stocktake (GST), Enhanced Transparency Framework (ETF), and the Santiago Network for Loss and Damage as well as the newly established process to establish funding arrangement to respond to Loss and Damage. These processes are selected because they each explicitly include Loss and Damage considerations and would require community-level information to represent the reality on the ground and policy conditions adequately. Processes with implicit inclusion or those yet to be decided (such as the New Collective Quantified Goal on Climate Finance) have not been assessed at this stage, but would remain relevant. These could, for instance, include National Adaptation Plans, and Action for Climate Empowerment.

3 Results and Discussion

3.1 The Steps of the Handbook for Community-Led Assessment of Climate-Induced Loss and Damage

The Handbook for Community-Led Assessment of Climate-Induced Loss and Damage (Anderson and Singh 2020) is formed of seven key steps and tools. Steps 1–3 collate data and are as follows: (1) "Mapping of Risks and Resources" to facilitate an understanding of the local-level geophysical changes resulting from climate change; (2) "Seasonal, Agricultural and Livelihood Calendars" to draw attention to how seasonal weather patterns are being affected by climate change, as well as the effects on farming, fishing, and livelihoods, and how these effects differ for women, men, and members of marginalised communities; (3) "Hazard Risk Index" to determine community and family infrastructure vulnerable to climate impacts.

Steps 4 and 5 are analytical. They are as follows: (4) "Trend Analysis" to enable intersectional insights and understanding about climate change impacts; (5) "Key Informant Interviews (KIIs)" which share the gathered realities with external experts. The international context and definitions of Loss and Damage are shared and unpacked in local realities.

Step 6 relates to "Calculating and Reporting Loss and Damage" by gathering the qualitative understanding of the climate change collective established by the community to develop a quantitative assessment of the cost of Loss and Damage.

The final Step 7 is "Advocacy and Lobbying". This critical step empowers the community to take the information and evidence to the relevant duty-bearers across levels.

3.2 Assessing the Handbook for Community-Led Assessment of Climate-Induced Loss and Damage to International Climate Policy Processes Under the UNFCCC and Its Paris Agreement

Each of the aforementioned 7 steps of the Handbook are relevant at the global policy space of the UNFCCC, its Paris Agreement, and other UN bodies. We keep in mind that this is also a way to capture indigenous knowledge especially when navigating local realities of Loss and Damage. This can be captured in the Holy See Secretary of State's statement during the High-Level Segment for Heads of State and Government at COP27: "We should also not neglect the non-economic side of Loss and Damage, like loss of heritage and cultures. Here, we have a lot to learn from indigenous peoples" (CAFOD 2022).

3.2.1 *Nationally Determined Contributions (NDCs)*

Article 3 of the Paris Agreement is the foundation of its structure laying out the approach of NDCs to allow all Parties to determine at the national level what actions they are able and willing to take to achieve the purpose of the Agreement. NDCs are determined by Parties and are not subject to negotiation.

At COP26, the decision text known as the Glasgow Pact (UNFCCC 2021) recognised clearly that the current NDCs are not ambitious enough. As the main vehicle through which to communicate ambition, Parties are to submit enhanced NDCs; however, currently, there is no framework in which to reflect Loss and Damage in NDCs. Despite this, an increasing number of countries have included Loss in their enhanced NDCs (Ryder and Calliari 2021), including Vanuatu (Republic of Vanuatu 2022), whose recent pioneering NDC details costs of addressing Loss and Damage as well as intended measures.

Whilst there is no formal guidance on how to include Loss and Damage in NDCs, Chamling Rai and Acharya (2020) provide four clear steps through which to integrate Loss and Damage into climate plans. These are: (1) Defining Loss and Damage in the national context; (2) Describing current and projected Loss and Damage; (3) Highlighting ongoing response to Loss and Damage; and (4) Including specific targets on Loss and Damage.

The Handbook has wide applicability in facilitating community-driven assessment of the Loss and Damage needs on the ground and reflects this back to the global community to signal the measures through which Loss and Damage can be addressed and the finance needs to do so. Utilising Chamling Rai and Acharya's four-step process, we can see how facilitative the 7 steps of the Handbook would be to enable the needs-based inclusion of Loss and Damage in NDCs. All 7 steps would be relevant and can be used across the other relevant processes below whilst providing a parameter through which to assess progress and signal the depth of funding needs in a comprehensive manner.

3.2.2 *Enhanced Transparency Framework (ETF)*

Article 13 of the Paris Agreement established another critical backbone of the agreement, which relates to the transparency of action and support.

It is an accountability mechanism, taking into account differing starting points and capabilities. It would take one or two major emitters not to act for the whole agreement to fail. There is no mechanism to "punish" countries if they do not act on pledges. Essentially, the Paris Agreement is a prisoner's dilemma built on trust with collaboration and competition as incentives. Reinforcement is needed from other major emitters to ensure that they are following suit.

The Enhanced Transparency Framework offers modalities, procedures, and guidelines that include a voluntary provision to report information related to Loss and Damage (UNFCCC 2022b). The ETF is relevant to Loss and Damage as every two years countries will report what they have done to "provide a clear understanding of climate change action" (Article 13.5 of the Paris Agreement) and "clarity on support provided and received" (Article 13.6). These should then lead towards the 5-yearly Global Stocktake.

Steps 1 through 7 of the Handbook are relevant to the ETF. Generally, most impact data is collected through existing monitoring mechanisms, focusing on economic losses. However, the ETF can be a route through which to showcase non-economic losses. This can also be shown through case studies which can be collated with the help of the Handbook.

3.2.3 Global Stocktake (GST)

Article 14 of the Paris Agreement refers to the GST. It brings up the critical ratcheting mechanism by taking stock of the implementation of the Paris Agreement to assess the collective progress towards achieving its purpose and long-term goals in a comprehensive manner. The GST will take place in 2023 at COP28 however leading to it is informed by a consultative process of technical dialogues and submissions. As Loss and Damage is enshrined under Article 8 of the Paris Agreement, the GST must assess the progress on averting, minimising, and addressing loss and damage.

At COP27, the GST transpired in World Cafe discussions. On the issue of Loss and damage, climate justice and inclusion were themes that arose and the need for a wide evidence base to track progress, from the global to the national and local. South Africa, for instance, stressed the need to go to the community level and incorporate indigenous knowledge solutions to formulate tangible solutions.

All 7 steps of the Handbook are relevant to the GST. It also offers a novel circularity of community-driven assessment of the ratcheting of the Paris Agreement. Climate Action Network's (2022) submission for the first input phase of the GST notes that Parties should provide evidence on how they: (1) Are planning on strengthening effectiveness of universal access to gender-responsive social protection; (2) Enabling, facilitating, and financing safe, orderly, and dignified movement of climate change displaced people and communities; (3) Incorporating participatory approaches for effective strategic planning to address and respond to L&D; and (4) Addressing L&D.

3.2.4 Santiago Network for Loss and Damage (SNLD)

The Santiago Network for Loss and Damage was established at COP25 to catalyse demand-driven technical assistance of relevant organisations, bodies, networks, and experts, for the implementation of approaches at the local, national, and regional levels.

An outstanding issue to resolve at COP27 in relation to the SNLD was the decision of the Advisory Board to operationalise the network. During the negotiations, it became clear that both developed and developing countries agreed on the need for a bottom-up approach to ensure the relevance of the SNLD. For instance, Canada noted on the composition of the Advisory Board: "it should be lean and representative of those of the frontlines" whilst the Africa Group of Negotiators stressed the need for a bottom-up approach to ensure country ownership and relevance.

The functions of the SNLD include: (a) catalysing demand-driven technical assistance, will need to be informed by what communities are demanding; (b) facilitating and catalysing collaboration, coordination, coherence, and synergies to accelerate action by organisations, bodies, and networks, and delivering effective and efficient technical assistance to developing countries; (c) facilitating the development, provision, and dissemination of, and access to, knowledge and information on averting, minimising, and addressing Loss and Damage across levels; (d) and facilitating access to action and support (UNFCCC 2019). For these functions to be relevant, they ought to reflect the lived experience of the vulnerable communities it is to serve. A fit-for-purpose SNLD would therefore be informed by real experiences of Loss and Damage to offer real solutions that can respond to communities' needs. As such, to deliver on these functions, the SNLD would require an approach that is inclusive and transparent, allowing for meaningful participation from frontline communities. Steps 1 through 5, of the Handbook are directly relevant to the political and technical space of the Santiago Network. It would empower locally led and community-driven needs assessment to ensure that any technical assistance is fit for purpose and context-relevant.

3.2.5 *Addressing Loss and Damage—The Relevance of the Handbook to the Question of a Loss and Damage Fund*

At COP26, the Glasgow Climate Pact mandated the Glasgow Dialogue on Loss and Damage. This was a substantial compromise for the world's biggest negotiating bloc, the G77 and China who were seeking to establish a finance facility or fund to address Loss and Damage.

During the first mandated Glasgow Dialogue in the UNFCCC Subsidiary Body intersessional period of June 2022, the conversations were clearly encapsulated in a 'North' 'South' divide. Whilst the outcomes of the Glasgow Dialogue were not officially captured, Civil Society was able to participate, listen, and contribute to the dialogue. Developing countries in their negotiating blocs and representing the majority of the world's population had for the most part clear needs-based demands that focused on the gap in finance to address Loss and Damage and the need to urgently close that gap. Developed country perspectives on the other hand were divorced from the local realities of impacted communities. This was reflected clearly in their argumentation and reluctance to accept that gap in finance to address Loss and Damage, which some developed countries and blocs such as the EU, the US, and the UK expressed as covered by existing modalities of finance such as for instance humanitarian aid, and adaptation finance.

Some Non-Party Stakeholders were able to intervene and highlight the perspectives of impacted communities. They presented a vision of what rights-based and justice-oriented finance to address Loss and Damage would look like. This perspective was widely espoused by developing country blocs and civil society alike.

Under this perspective, such finance would be based on the cooperation and solidarity of the developed countries to the developing countries through the provision of finance to address Loss and Damage. This could be achieved by setting up a Loss and Damage Fund adhering to the principles of (1) international solidarity, historical responsibility, and the polluter pays principle; (2) new and additional; (3) needs-based, adequate, predictable, and precautionary; (4) locally driven with subsidiarity, enveloping gender responsiveness and equitable representation; (5) public and grant-based; (6) balanced and comprehensive.

During COP27, an agenda item was agreed to discuss Loss and Damage finance for an early historic first. The item arose under the COP agenda item 8 and CMA agenda item 8: Matters relating to finance, sub-agenda item (f) Matters relating to funding arrangements responding to Loss and Damage associated with the adverse effects of climate change, including a focus on addressing Loss and Damage.

The negotiations under 8(f) could be grouped into two stances. One on side there was a united G77 and China position on their expected outcome under 8(f). Specifically, the bloc's expectation was for 8(f) to yield a decision to establish a Fund to address Loss and Damage as an operating entity under the Financial Mechanism of the UNFCCC (G77 and China 2022). Meanwhile, developed country positions vied away from such a fund offering and sought solutions such as gap analysis of existing arrangements. This can be evidenced by statements during the negotiations of this agenda item by Northern blocs such as the EU: "at COP28 we can take some decisions where gaps and funding arrangements are. This could be the conclusion of final outcomes in 2024".

However, during the negotiations under the agenda item, the developing countries made it clear that this was not the case. For instance, Ecuador stated: "We are very adamant and insistent on Loss and Damage needs—we will have to make those needs evident (not in biennial reports)". Maldives gave a real-world example: "We are at the brink of climate impacts. Addressing Loss and Damage for me is to continue to live a healthy life in the Maldives. Usually when something happens, a flood of organisations come to deal with immediate needs and then they leave but we need something to respond to slow-onset impacts". India further stressed: "to put on record that mitigation is averting, minimising is adaptation, here you are talking about addressing Loss and Damage, it is a real impact. Humanitarian aid is not [addressing] Loss and Damage. Addressing Loss and Damage is about resilient recovery and reconstruction after a disaster by building back better. Livelihood loss, and also non-economic losses and scaling-up actions and support on Loss and Damage, is an immediate and urgent need".

The subsequent decision text (UNFCCC 2022a) acknowledged that there was a gap in addressing Loss and Damage and established a Transitional Committee (TC) through which to operationalise new funding arrangements for assisting the developing countries that are particularly vulnerable to the adverse effects of climate change, in responding to Loss and Damage, including with a focus on addressing Loss and Damage. It was also decided that the TC will be informed of the gaps within that current landscape.

The steps of the Handbook are particularly relevant to offering a holistic assessment of the gap in finance to address Loss and Damage. The data collected in this manner can lend argumentation and evidence to the question of gaps and needs directly from those whom the fund is meant to serve. It can offer critical needs-based and intersectional insights whilst also guiding and facilitating locally informed access modalities to ensure that finance flows to the communities that need it. As such the application of the Handbook can be a key in developing indicators in the establishment of fit-for-purpose fund to address Loss and Damage.

The value of assessments carried out by local groups and people who have first-hand experience with the effects of climate change to their lived experiences can facilitate the development of a truly bottom-up fund which addresses the needs of the vulnerable. For policymakers, such information can help determine the categories of Loss and Damage most common in certain communities and create focused and efficient finance strategies to mitigate these effects.

Ultimately, the assessments as carried out via the steps of the Handbook can aid in making sure money is allocated to the neediest populations. Many of the populations most at risk from climate change's effects are also the ones least equipped to speak up for their own needs but this can equip them with an outlet to be included in ultimately a platform that is after all meant to serve them.

4 Discussion

The Handbook is relevant to international climate policy processes under the UNFCCC and its Paris Agreement because it provides a tool for communities to understand and communicate how climate change is affecting their lives and livelihoods. In doing so, it empowers communities to take hold of their knowledge and the valuable data of their lived experience of climate change. Governments and other stakeholders can use this data to inform the creation and execution of climate policies and actions, including Nationally Determined Contributions (NDCs).

Communities can use the Handbook to pinpoint their vulnerabilities and needs whilst also creating plans to lessen the impact of and their exposure to climate hazards. Policymakers and other decision-makers can use this information to develop more effective and targeted policies and activities to address the local impacts of climate change. The Handbook can also aid in increasing communities' capacity to engage in climate policy discussions and advocate for the inclusion of their needs and top priorities.

Community-led Loss and Damage assessments can be useful for the identified process of the UNFCCC and its Paris Agreement in several ways including:

1. Providing a more comprehensive and accurate understanding of the impacts of climate change on communities. Community-led assessments are often conducted by local organisations and individuals who have first-hand knowledge of the impacts of climate change on their communities and can provide valuable insights and information that may not be captured in more formal, top-down assessments.
2. Helping to ensure that the needs and priorities of the most vulnerable communities are considered.
3. Enabling local capacity and resilience. Participating in the assessment process can help communities to better understand the risks and impacts of climate change, and to identify ways to adapt and become more resilient. This can help build communities' capacity to better withstand future climate-related shocks.
4. Promoting greater transparency and accountability by involving communities in the assessment process. Policymakers can demonstrate their commitment to transparency and accountability and build trust with local communities. This can be particularly important in contexts where communities may have historically been marginalised or excluded from decision-making processes.

In addition, in the case of a continued absence of cooperation and solidarity in relation to finance and support for addressing Loss and Damage, the Handbook offers the tools to collate evidence which can be facilitative to communities seeking justice through other means such as litigation. It would be complementary to assembling evidence for policy instruments such as the liability roadmap (www.liabilityroadmap.org) which shows how to employ a set of tools to hold polluters liable for Loss and Damage incurred.

At the local level, for instance, we see individuals and communities engaging legal levers to hold historical polluters accountable by taking legal and legislative actions:

- Saúl versus RWE: in 2015 a Peruvian farmer and mountain guide filed a lawsuit against the German utility RWE citing the threat to family, property, and the city of Huaraz from the massive emissions of the company (Germanwatch 2022).
- The 2022 example of legal recognition that the obligation of states to uphold human rights was violated or failed—and shown in the U.N. Human Rights Committee (OHCHR 2022) ruling which found that Australia's failure to adequately protect indigenous Torres Islanders against adverse impacts of climate change violated their rights to enjoy their culture and be free from arbitrary interferences with their private life, family, and home.

By building the evidence base of the rights, justice, and equity landscape, communities can point to the legal obligations and fundamental rights which are being infringed upon and take to task the multilateral process and their national governments.

5 Conclusion

Participatory approaches are necessary for effective strategic planning and ownership to respond to and address climate-induced Loss and Damage. As described in the Handbook, evidence gathering can strengthen and clarify the linkages between climate change impacts, human rights, and lived realities of affected communities. This can strengthen the effectiveness of international response and the vectors of support provided at the scale commensurate to the needs.

Overall, the Handbook for Community-Led Assessment of Climate-Induced Loss and Damage is an important tool for supporting the participation of communities in international climate policy processes and for ensuring that the needs and perspectives of these communities are considered.

Next steps would be to work with community leaders in the rolling out of this analysis will be an essential data basis for future work in the national and multilateral policy space. In the near term, the analysis will be fundamental for the identified processes under the UNFCCC and its Paris Agreement with urgency for the critical conversations that will be taking place through 2023. Namely, these include the evidence basis for the Global Stocktake bringing in assessments through the NDCs, and ETF which can convey the scale and dimensions of needs of affected communities. In addition, ensuring suitable SNLD and Loss and Damage funding arrangements facilitated by locally led and community-driven needs assessments to ensure that any technical and financial assistance is fit for purpose and context-relevant.

Acknowledgements The authors acknowledge the support of Asia-Pacific Network for Global Change Research (APN) for supporting the development of the Handbook in the first instance and for the opportunity to deliver this paper.

We extend our appreciation to Dr Senaka Basnayake and Hina Lotia for their time and patience in reviewing this paper.

Finally, we extend gratitude to ActionAid and Teresa Anderson for their vital work in the development of the Handbook as well as the community members—women, men, children and young people—without whose participation the Handbook would not have been possible.

References

Anderson, T., & Singh, H. (2020). Participatory methodologies enable communities to assess climate-induced loss and damage. *APN Science Bulletin*, 10 (1), 132. Retrieved from: https://www.apn-gcr.org/wp-content/uploads/2021/03/APN-Science-Bulletin-2020-compressed.pdf#page=134

AOSIS (2022, June 16). *SBSTA SBI Closing Statement*. Retrieved from: https://www.aosis.org/progress-is-inadequate-sbsta-and-sbi-closing-plenary-sb56-statement/

CAFOD (2022). *Message from His Eminence Cardinal Pietro Parolin OMRI, Holy See Secretary of State, to the COP27 leaders' summit*. Retrieved from: https://cafod.org.uk/News/Campaigning-news/Vatican-COP27

CAN International (2022). *CAN submission for the first input phase of the Global Stocktake*. Retrieved from: https://climatenetwork.org/wp-content/uploads/2022/08/GST-Submission-Climate-Action-Network-August-2022.pdf

Chamling Rai, S., & Acharya, S. (2020). Anchoring loss & damage in enhanced NDCs. *WWF and Practical Action*. Retrieved from: https://wwfint.awsassets.panda.org/downloads/wwf_pa_anchoring_loss_and_damage_in_ndcs_report.pdf

G77 and China (2022, November 15). *COP27 8(f) Submission—Matters related to funding arrangements for responding to loss and damage, including a focus on addressing*. Retrieved from: https://unfccc.int/event/cop-27?item=8%20f

Germanwatch (2022). *The Climate Case: Saúl versus RWE*. Retrieved from: https://www.germanwatch.org/en/huaraz

Lo, J., & Farand, C. (2021, November 3). Scotland breaks loss and damage "taboo", raising hopes others will follow. *Climate Home News*. Retrieved from: https://www.climatechangenews.com/2021/11/03/scotland-breaks-loss-damage-taboo-raising-hopes-others-will-follow/

OHCHR (2022). *Australia violated Torres Strait Islanders' rights to enjoy culture and family life, UN Committee finds*. Retrieved from: https://www.ohchr.org/en/press-releases/2022/09/australia-violated-torres-strait-islanders-rights-enjoy-culture-and-family

Pörtner, H.-O., Roberts, D. C., Adams, H., Adler, C., Aldunce, P., Ali, E., Begum, R. A., et al. (2022). *Climate change 2022: Impacts, adaptation, and vulnerability. Contribution of Working Group II to the Sixth Assessment Report of the Intergovernmental Panel on Climate Change*.

Republic of Vanuatu (2022). *Vanuatu's revised and enhanced 1st Nationally Determined Contribution 2021–2030*. Retrieved from: https://unfccc.int/documents/578782

Ryder, B., & Calliari, E. (2021). How does Loss and Damage feature in Nationally Determined Contributions? *CCLAD, UCL*. Retrieved from: https://www.ucl.ac.uk/global-governance/sites/global_governance/files/ld_ndc_pb.pdf

Santos, S., Ravara, P., & Rodrigues, D. (2021). Guide for participation in democratic environmental governance. *WFD*. Retrieved from: https://www.wfd.org/sites/default/files/2022-02/Env.Democracy%2006.05.21.pdf

Singh, H. (2021, December 10). Polluters risk litigation if they fail to address loss and damage. *Third Pole Network*. Retrieved from: https://www.thethirdpole.net/en/climate/loss-and-damage-polluters-risk-litigation/

UNFCCC (2019). *Agenda item 7 Warsaw International Mechanism for Loss and Damage associated with Climate Change Impacts*. FCCC/PA/CMA/2021/L.22. Retrieved from: https://unfccc.int/sites/default/files/resource/cma2021_L02E.pdf

UNFCCC (2021). *The Glasgow Pact*. 1/CP.26. Retrieved from: https://unfccc.int/documents/310475

UNFCCC (2022a). *Funding arrangements for responding to loss and damage associated with the adverse effects of climate change, including a focus on addressing loss and damage*. Decision 1/CP.27 1/CMA.4. Retrieved from: https://unfccc.int/documents/624440

UNFCCC (2022b). *Reference manual for the Enhanced Transparency Framework under the Paris Agreement—Understanding the enhanced transparency framework and its linkages*. Retrieved from: https://unfccc.int/sites/default/files/resource/ETFReferenceManual.pdf

CHAPTER 4

Pillars of Resilience: Assessing Loss and Damage at the Local Level

Hina Lotia

Abstract The chapter discusses the emergence of 'Loss and Damage' at COP13 in 2007 and the subsequent adoption of various mechanisms to address it in vulnerable developing countries. The chapter presents a methods toolbox for the local-level assessment of Loss and Damage resulting from climate-related stressors, which was applied in three South Asian countries. It identifies six key elements of local resilience that must be integrated into rural adaptation planning. It includes impacts on livestock, agriculture, human health, micro-enterprises, migration, and infrastructure. The study explores the linkages of these elements of resilience with each other to demonstrate the importance of an integrated approach. Additionally, the chapter reviews Pakistan's national and provincial climate policies and action plans to analyse if the resilience elements are addressed. The objective is to provide recommendations

H. Lotia (✉)
The World Bank/Ministry of Planning, Development and Special Initiatives, Islamabad, Pakistan
e-mail: hlotia@gmail.com

V. Wijenayake et al. (eds.), *Linking Climate Change Adaptation, Disaster Risk Reduction, and Loss & Damage*,
https://doi.org/10.1007/978-981-99-8055-0_4

for the proposed National Adaptation Plan and possible development of policies and strategies to include Loss and Damage considerations.

Keywords Loss and Damage · Assessment toolbox · Climate change adaptation · Disaster risk reduction · Pakistan

Highlights

- Establishment of a Loss and Damage fund at COP27 has spurred the discussion on the modalities and mechanisms to form the basis for financing.
- No agreed methodology to assess Loss and Damage exists.
- Adaptation has limits, and beyond that, there is a cost to building resilience.
- Extreme events are pushing limits for vulnerable population, driving them further down the poverty ladder.
- Need for policies and strategies to aptly reflect various elements of Loss and Damage.

1 Introduction

Climate change and its impacts have been attributed to measurable and non-measurable costs to both public and private goods. According to a study, from 1970 to 2019, there were 22,326 disasters, leading to 4,607,671 deaths and US$ 4.92 trillion in economic losses. Of these disasters, 11,072 have been attributed to weather, climate, and water hazards, resulting in 2.06 million deaths and US$ 3.64 trillion in losses (WMO 2021). Empirical knowledge emanating from research, including the costs of such damages associated with climate change, are important indicators used by policymakers and decision-makers to understand and comprehend the social and developmental challenges. These may be exacerbated by the exposure to one or more climatic hazards, indicating the level of sensitivity of systems under study. Such information, if synthesized, plays an important role in helping communities prepare for and minimize these effects. The adverse impact of climate change on humans, society, natural, and built environment is referred to as Loss and Damage. Intensity, frequency, and geographical distribution of extreme events, such as floods, droughts, sea-level rise, and desertification are the manifestation of climate change (WMO 2021). These result in Loss and Damage, classified as economic (WMO 2021) and non-economic (Grantham Research Institute on Climate Change and the Environment 2022).

Loss and Damage is an emerging topic in the climate change arena. When mitigation and adaptation, as well as preparedness and Disaster Risk Reduction measures undertaken, are not optimally implemented or where limits to adaptation are reached to avoid the impacts on an ecosystem or a community, Loss and Damage is incurred. *Loss* refers to permanent losses or losses that cannot be recovered, while *Damage* refers to impacts that may be reversed at a cost. It is often the case when people, community or governments cannot take actions as they are not affordable, technically, physically feasible or socially challenging. Policymakers, particularly in the most vulnerable countries, grapple with how to address current and future climate-related losses and damages.

1.1 Genesis of Loss and Damage in International Climate Negotiations

Loss and Damage was first introduced in the UN climate negotiations in 1991 by the Alliance of Small Island States (AOSIS) for countries that would be impacted by sea-level rise and would lose their land. The term Loss and Damage was referred to in the United Nations Framework Convention on Climate Change (UNFCCC) texts in the 2007 international climate negotiations in Bali (Indonesia) during COP13. In subsequent COPs, Parties acknowledged the efforts to address the Loss and Damage associated with the adverse effects of climate change, stressing the need to scale up efforts by enhancing support and coordination in the broader context of climate-resilient sustainable development (UNFCCC 2012).

In 2013, Loss and Damage gained momentum when Parties to the UNFCCC in COP19 agreed to establish the 'Warsaw International Mechanism for Loss and Damage' associated with Climate Change Impacts (WIM). The mechanism is meant to 'facilitate dialogue for enhanced knowledge, coherence, action and support to avert, minimize and address Loss and Damage associated with climate change impacts in developing countries that are particularly vulnerable to the adverse effects of climate change'. Article 8 of the Paris Agreement at COP21 also stressed that Parties should enhance understanding, action, and support on a cooperative and facilitative basis with respect to Loss and Damage.

In 2019, countries agreed to establish the 'Santiago Network' on Loss and Damage at COP25 in Madrid. It aimed to connect developing countries with providers of technical assistance. Subsequently, in the 'Glasgow Climate Pact' at COP26 in 2021, the Parties agreed to establish a dialogue 'to discuss the arrangements for the funding of activities to avert, minimize and address Loss and Damage associated with the adverse impacts of climate change'.

At COP27 in Sharm el-Sheikh in 2022, Loss and Damage finance was included in the formal agenda and Parties agreed to establish a Loss and Damage Fund. Setting up a fund was an achievement that the countries vulnerable to the impact of climate change have been working on for many years. However, many decisions, including the fund's placement, types of activities to be supported, eligibility criteria for support, and financial contribution to the fund, remain to be sorted. The work of the 'Transitional Committee' to give its recommendations at COP28 in 2023 becomes more meaningful as these critical decisions await resolution.[1]

1.2 Enabling Factors for an Effective Loss and Damage Fund

Despite progress in setting up a dedicated fund for Loss and Damage, the question of agreeing on a universal approach, method, and tool to assess the risk of and to respond to Loss and Damage associated with the adverse effects of climate change will become paramount. Regional, national, and local capacity, context, and circumstances will be paramount. Assessing Loss and Damage, or for that matter, the risk of Loss and Damage associated with the adverse effects of climate change, including slow-onset impacts, will need to be defined. The need for systematic observation of, and data collection on, the impacts of climate change, particularly slow-onset impacts, and accounting for losses, as appropriate would need to be streamlined. Involvement of vulnerable communities and populations, and civil society, the private sector, and other relevant stakeholders, in the assessment of and response to Loss and Damage would be crucial. Work to advance the understanding of and expertise on Loss and Damage,

[1] https://www.pbs.gov.pk/sites/default/files/population/2017/tables/punjab/Table01p.pdf.

including economic and non-economic losses and damages, and how Loss and Damage associated with the adverse effects of climate change affect those segments of the population that are already vulnerable owing to geography, gender, age, indigenous or minority status, or disability, needs benchmarks.[2]

With substantial development with reference to a financing facility on Loss and Damage, there is a need to consolidate the work done on understanding and assessing Loss and Damage under various climatic conditions. What works in a certain geographical location, with a particular climate stressor, in a deprived rural setting in a developing country, may not be suitable for an urban locale with completely different socio-economic conditions and vulnerability. The work on assessment of Loss and Damage becomes more relevant when the functions, modalities, and operational mechanism will be devised to determine financing needs.

To address Loss and Damage adequately, a robust methodology to assess Loss and Damage is needed for evidence-based policy decisions. The work funded by APN focused on developing a methods toolbox to become a key instrument and reference for future studies on Loss and Damage. It was designed for a wide range of settings, in areas with different climate-related stressors (floods, droughts, cyclones, sea-level rise, glacial melt, coastal erosion, changing monsoon patterns, etc.) and different levels of socio-economic development (though designed primarily for use in developing countries).

As a case study, Rajanpur district located in the far south-west of Punjab, Pakistan was selected to apply of the toolkit, and to assess Loss and Damage from recurrent floods. The district lies between the Indus River on the east and the Suleiman Mountains on the west. It spans a geographical area of 12,318 sq. km with a population density of 162.04 sq. km and 85.5% of the district population resides in rural areas whereas the remaining in urban areas.[3] The district's climate is dry and annual rainfall is unpredictable and low (less than 15 mm). The soils are sandy, loam-clay, and loam. Irrigation relies on floods of the hill torrents.[4] With a population of 1,103,618 people (census 2017), with 52.6% of male

[2] https://www.fao.org/3/y5460e/y5460e06.htm.

[3] https://www.pbs.gov.pk/sites/default/files/population/2017/tables/punjab/Table01p.pdf.

[4] https://www.fao.org/3/y5460e/y5460e06.htm.

and 47.4% female.[5] With an average annual growth rate (1981–1998) of around 3.3%, only 8% of the households have piped water supply.[6] It is classified as a rural area, where housing structures are mostly Kuchha (only 17% of the houses are made of concrete—'Pacca'), and only 40% of population has access to electricity.[7] In almost all human development indicators, the area is below national averages and falls in the bottom 25 per cent of districts.

The manuscript focuses on the case study's findings and underscores the importance of six key elements for assesing of Loss and Damage in the rural setting of Rajanpur. It will highlight the vulnerabilities of the poor and the marginalized, the measures undertaken to build resilience, and the limits to adaptation. It will also underline the importance of including certain elements as part of the policy and planning—to assess Loss and Damage and to access climate finance from the Loss and Damage funding facility when it becomes operational. The recommendations will contribute towards the body of knowledge and will possibly also serve as input for the Transition Committee in their work.

2 Methodology

The methods used in this draft combine quantitative and qualitative approaches. Based on an extensive literature review, a methods toolbox for local-level assessment of Loss and Damage from climate-related stressors, including sudden-onset events and slow-onset processes, was developed. An intensive workshop was held for researchers and experts to refine the toolkit as well as the associated data collection forms. Enumerators from community-based organizations were trained on the use of the questionnaire. The training included detailed sessions on the conceptual framework of Loss and Damage and explanations of the data collection objectives. Sample questionnaires were filled out during the training, and the results were discussed. Pilot data collection was carried out to test the questionnaire and the sampling methodology in the field.

Primary data was collected from households to assess the effects of Loss and Damage that were assisted and supervised in the field along with

[5] https://www.pbs.gov.pk/sites/default/files/tables/district_at_glance/Rajanpur.pdf.

[6] https://www.pbs.gov.pk/dag-punjab.

[7] https://www.pbs.gov.pk/sites/default/files/tables/district_at_glance/Rajanpur.pdf.

the enumerators. Primary data collection was augmented by Focus Group Discussions (FGDs) with community members and expert interviews with government representatives.

The results from the case study were analysed in conjunction with international developments on Loss and Damage. Key elements were underlined from the Rajanpur case study to put forth a strong narrative for inclusion in policies and strategies at the national as well as international level. Climate change policies at the national and provincial levels in Pakistan were also reviewed to see if Loss and Damage in general and the key elements, in particular, had been adequately covered.

3 Results and Discussion

The work funded by APN focused on developing of a methods toolbox for local-level assessment of Loss and Damage from climate-related stressors, including sudden-onset events and slow-onset processes. Conceptually and methodologically, the toolbox combined climate change adaptation and disaster risk reduction perspectives, paying close attention to adaptation limits and constraints as well as individual risk management strategies that people adopt to prevent or minimize disaster losses. The methods toolbox was developed as a key instrument and reference for future studies on Loss and Damage, with experiences from multi-country studies on Loss and Damage done from the perspective of affected people in the least developed and other vulnerable countries, including three in Asia. It was designed for application in a wide range of settings, in areas with different climate-related stressors and varied levels of socio-economic development (though designed primarily for use in developing countries).

3.1 Learning from Rajanpur Case Study

The toolkit was applied in district Rajanpur as a case study in 2015, which has a long history of floods. Monsoon floods in 2006 and hill torrents during 2005, 2007, and 2008 affected most of the population. In 2008, the overflow in the Indus River caused heavy flooding at Jampur, Rajanpur, and Rojhan tehsils affecting 62 villages (Nadeem et al. 2014). In 2010, severe flash floods were recorded in Rajanpur (Nadeem et al. 2014) that destroyed cropped areas and considerable damage to livestock was also reported. The floods of September 2012 caused widespread loss of life, livelihoods, and infrastructure in Rajanpur. Torrential rains in

2013 gave rise to heavy flash floods, demolishing houses, lost possessions, fields inundated; and link roads destroyed. 187,000 acres of farmland were swept away and over 165,000 people were affected.[8] The rainfall patterns in Rajanpur have become erratic, making it difficult for communities to predict and rely on local rainfall patterns.

The residents of the areas interviewed for the case study in Rajanpur district survive on subsistence agriculture as they are at a distance from markets and, for the most part, mainstream economy. Their livelihood options are narrow, limited to small-scale agriculture and livestock. Their physical assets are limited and have very small cash or market value. Their resources for autonomous and planned adaptation are, therefore, extremely limited as are the government interventions in the area to augment their coping capacity for extreme weather events. Each flooding episode diminishes their resilience and enhances their vulnerability to the next flooding, the frequency and scale of which has been increasing in the area.

3.2 *Pillars of Resilience*

The case study highlighted that the climate change adaptation in Rajanpur is limited to only a few Union Councils[9] and the understanding of climate change impacts and vulnerabilities is limited. Communities have developed their own adaptation and coping mechanisms to deal with climate change's unpredictable and devastating effects. The effectiveness of household preventive and coping measures were documented. Six key elements were drawn upon from the case study where coping mechanisms fall short of resilience and have implications for discussion on Loss and Damage valuation, compensation, and relief. These include impact on (i) agriculture, (ii) livestock, (iii) micro-enterprises or sources of non-farm income, (iv) human health, (v) infrastructure, and (vi) migration (Fig. 1).

The six elements serve as pillars of household income and security, climate vulnerability and resilience. Since their individual, family, and communal lives are within the sphere of these elements, their autonomous and planned adaptation rests on these pillars.

[8] https://www.emro.who.int/images/stories/pakistan/documents/who-health_sitrep__1_05_08_2013_-final.pdf?ua=1.

[9] Union Council is the third tier of governance arrangement, after 'District' and Tehsil being the first and second tier under a Province.

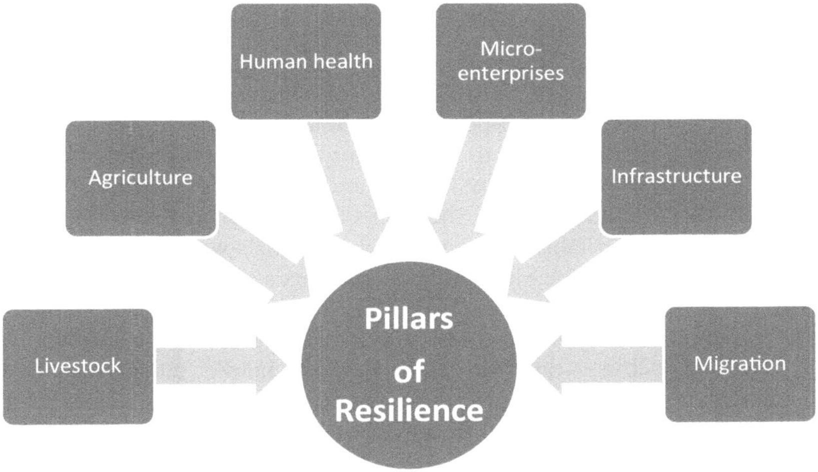

Fig. 1 The pillars of resilience

3.3 *Agriculture*

In Rajanpur, agriculture is the source of livelihood for most people. In terms of adaptive measures, bunds were constructed, retention walls were built, and climate-resilient seed varieties were sown. The intensity and ferocity of each successive floods in the areas led to losses and damages beyond the adaptive capacity. Frequent floods not only damaged their crops but also reduced soil productivity. As an adaptive measure, the households store dried wheat, vegetables, and pickles in order to secure food supplies during floods. However, it was reported that these stored food supplies either got washed away or were ruined due to poor food storage facilities.

Further, floods disrupted not only regular farming activities but also the marketing of the produce as they could not always reach markets where the crops were sold. In the absence of government intervention and corrective market mechanisms, for farmers, the markets operate at a lower price than the market price. The ability of farmers to cope with the impacts of floods was further diminished, with far-reaching impacts on food security for the entire country.

3.4 Livestock

In the case study areas, livestock rearing is practised for commercial and subsistence needs. They are saleable assets for farmers and are also used for ploughing and harvesting. Surplus dairy products and meat from livestock rearing are sold in the market to supplement household incomes. Though limited measures were taken by farmers and households, such as having animal sheds in raised locations, the intensity of floods adversely impacted livestock as their mortality rates increased as animals were washed away.

Floods damaged animal health, feed, and sheds and created a shortage of dry land for hosting livestock. The treatment of livestock is a significant expenditure that households cannot reclaimed, leading farmers to sell their livestock at a rate that is less than the market rate. As the loss of animals represents loss of financial assets, farmers suffered significant financial losses during and after floods. Loss of income from reduced livestock ownership has long-term impacts. High cost of machinery to replace livestock may not be available to most farmers. In addition, surplus meat and dairy products to supplement incomes diminish. The buffer against future shocks is reduced, animal-assisted crop cultivation and harvest are threatened, and it makes the farmers more vulnerable.

3.5 Micro-enterprise

Community members engage in micro-enterprises/non-agriculture activities to deal with the impacts of floods, mainly to buy food, clothing, and shelter, and to meet travel expenses for migration. The case study highlighted that with the agriculture land impacted by floods, job opportunities for labourers were reduced. The local labour market suffers during floods and the unskilled community-level workers are not fairly paid for their products and services, making them even more vulnerable. This encouraged migration to nearby areas, both urban and rural. Floods impacted small businesses, which were closed due to damage to infrastructure. With diminished financial resources, women could not continue their local crafts as the demand for such products was reduced.

3.6 Human Health

Floods negatively impact the local communities' physical and psychological health. Disease outbreaks in the aftermath of floods pose increased expenditures for treatment. In such situations, as they need to spend more on health, food, and nutrition, they have fewer resources and skewed income levels due to the floods.

In Rajanpur, loss of life was relatively limited but there was a disease outbreak, especially water-borne diseases posing increased household's expenditure.With limited resources available for food, malnutrition and stunting among newborns were evident. The mental health of many was affected, with people facing depression due to the complete loss of their homes, livelihood, cultural heritage sites, and disconnection from family and friends. Many women faced depression and anxiety as living in shelters was socially difficult.

3.7 Infrastructure

In the case study areas, houses in the village were made mainly of natural material with earth floors. Many of them did not have physical barriers for protection. They were entirely or partially damaged and household assets such as bikes, radios, utensils, etc. were also lost. The villagers would store food as a preventive measure, but the lack of proper storage facilities resulted in the food being destroyed, resulting in an exponential increase in food prices. Availability of drinking water also became an issue as people got drinking water from hand pumps and wells installed in the village.

As an adaptive measure, dykes and embankments were built along the riverbanks but they could not withstand the greater and sudden surge of water in the riverine areas. Measures to build houses on raised platforms were insufficient, as with the increased impact of climate change, the severity and frequency of flooding in Rajanpur also increased. The water levels were so high that they washed away entire household and livestock. Many households secured their possessions on trees, but they were stolen or washed away. Loss of irreplaceable belongings was an emotional and financial trauma.

3.8 *Migration*

In Rajanpur, community members either shifted to safer places (such as schools and mosques) or migrated to other cities to deal with the impact of floods. Costs incurred for migration and relocation cannot be regained. They represent a significant portion of the expenditure and may need to be financed through loans or asset liquidation. The emotional and psychological impacts of moving to a different city and the problems faced by community members therein are still unaccounted.

While migration was considered an autonomous coping mechanism to survive or mitigate impacts in the short term, it decreased the resilience of communities in the long term. It pushed them further down the poverty spiral and often they became worse-off. To rebuild their livelihood, many residents resort to loans, and extreme floods in subsequent years diminish their ability to repay, making them more vulnerable and less likely to get loans in the future.

3.9 *Challenges Preventing Resilience*

Two of these pillars (agriculture and livestock) have traditionally served as sources of their livelihood, sometimes supplemented by their micro-enterprises, serving as a source of additional or side income or as a backup to failed crops or losses in livestock. These two, together with their shelter/houses, serve as elements for their economic stability and, in good times, source of well-being. The government's declining investments in health infrastructure and otherwise increased frequency of disasters have posed several health-related challenges for which they often lack financial resources.

The case study has highlighted that vulnerable households face Loss and Damage when exposed to extreme events, nominal despite mitigation and adaptive measures undertaken in the area. According to the survey conducted, difficulty in adopting effective measures was not an issue. However, the main impediment was the lack of finances and resources to instal preventive measures. Often, skills and knowledge were a challenge, but the community members also felt that the government did not help enough to take preventive measures. Local non-governmental organizations (NGOs) did play a supporting role but could not adequately address the vulnerabilities and extent of the damage. The measures taken were inadequate to face the catastrophe resulting from climate change.

Borrowing has also become more frequent as the floods have become more frequent and many of them are forced to migrate seasonally to safety during the flooding season. The hexagonal interaction of these six pillars defines their adaptation capacities and limits. While the flood timing (in terms of cropping season) velocity, duration, and quantity of standing water, and frequency and periodicity determine the losses and damages caused by the floods, the six elements determine the level of their vulnerability and/or resilience.

3.10 Interlinkages Reinforcing Climate Vulnerability

The six pillars/elements of Loss and Damage identified in the case study and mentioned above are intrinsically linked. The case study shows that each one of these is a contributing factor for the other elements. Emerging evidence shows that possible income increases from livestock and agriculture enable community households to spend more on health. Inversely, reduced income from these two sources (livestock and agriculture) or losses during floods or crop failures can impact their microenterprises or non-agricultural incomes. Any income increase contributes to their efforts to invest in their shelter or house. This can be in terms of expanding to meet their needs or strengthening it from floods—the most frequent and disastrous extreme weather event. The house damaged and destroyed is often also accompanied by crop destruction and loss of livestock. These material losses diminish their resilience and they resort to migrating to safety with the remaining assets (Fig. 2).

These six pillars are effectively the six elements of their resilience. The weakening of one contributes to the weakening of the others. For those living in poverty, the accumulative value of each household's assets is extremely thin; therefore income from each source (livestock, agriculture, micro-enterprise) enhances the capacity only marginally to improve the house condition or undertake health-related expenditures. Against this background, the vulnerability to climate shocks is extremely high. The opportunities and interventions by local government institutions are extremely limited to enhance resilience.

Based on a globally recognized damage, loss and needs assessment methodology,[10] the Government of Pakistan led the Post-Disaster Needs

[10] Jointly developed by the EU, the World Bank Group (WBG), and the UN.

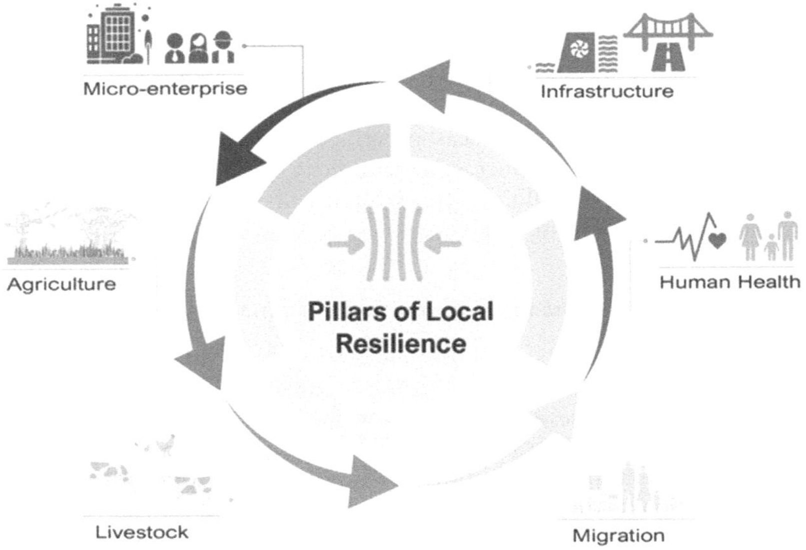

Fig. 2 Pillars of local resilience

Assessment (PDNA).[11] Focusing on the 94 calamity-hit districts across the four provinces of Balochistan, Khyber Pakhtunkhwa, Punjab, and Sindh, it analysed macroeconomic and human impacts and a summary of sector assessments of the 2022 floods.

The assessment estimates that extreme floods led to more than US$ 14.9 billion in total damages. The total economic loss was about US$ 15.2 billion and the resilient needs for rehabilitation and reconstruction are not less than US$ 16.3 billion. Any new investment needs to support the country's adaptation and resilience to climate change, or future climate shocks are not accounted in these calculations.[12] The PDNA reports specifically highlight that although Punjab province was largely spared from the floods, Rajanpur was one of the two districts damaged

[11] Led by the Government of Pakistan and supported by the Asian Development Bank, the European Union (EU), the United Nations (UN), and the World Bank.

[12] https://www.undp.org/pakistan/publications/pakistan-floods-2022-post-disaster-needs-assessment-pdna.

by the unprecedented deluge. Floods in 2022 and the heavy economic and social losses caused by them have washed away the local infrastructure, standing crops, and livestock. It has shown that the condition of the communities has not improved since the survey was conducted, visits were undertaken, and the case study was written. If anything, their resilience has eroded and their vulnerability has increased.

3.11 Policy Domain to Support Loss and Damage

While there is considerable evidence from the case study regarding the importance of building resilience under the six elements of Loss and Damage, there is a need to review various climate change policies and strategies at the national and provincial levels (Table 1).

The government does various surveys and assessments on a regular interval, but climate change, vulnerability, and resilience indicators are not systematic included as part of the survey questions. Population census, agriculture census, and labour force survey do not account for climate sensitivities. Similarly, a national risk-profiling exercise has not been carried out, making most public projects prone to climate shocks and vulnerabilities. Risk assessments are carried out on an ad hoc basis but do not collect data on all climate hotspots or have a standard methodology to carry out any review.

4 Conclusions

The six elements or pillars of resilience identified in the paper are closely linked to the geographical, economic, and social circumstances of the Rajanpur district of Punjab province but have equally clear relevance for other flood-prone areas of the country. Climate variability and the associated risks in the area magnify the vulnerability of each extreme event. The frequency and intensity of the subsequent disasters expose the lives and livelihood of the communities in risk-prone areas. These pillars help identify Loss and Damage faced by the poor and the vulnerable. In other geographical locations, additional elements may highlight the burden to recover from the impacts of climate crises.

The National Climate Change Policy needs to embed the three functions outlined in the Warsaw International Mechanism for Loss and

Table 1 Loss and damage in climate change policies and strategies

Climate change policies and strategies	*Analysis of loss and damage components*
National Climate Change Policy (2021)	The goal of the national policy is to ensure that climate change is mainstreamed in the economically and socially vulnerable sectors of the economy and to steer Pakistan towards climate compatible development. The policy refers to L&D under the adaptation policy measure for disaster preparedness It refers to development of an 'assessment and compensation mechanism' including insurance for L&D in the aftermath of disasters and measures for infrastructure and soil rehabilitation. In this context, the L&D assessment toolkit should be included as a policy tool to be adopted nationwide

Climate change policies and strategies	*Analysis of loss and damage components*
Nationally Determined Contributions (NDC 2021)	The NDC aims at achieving reduced poverty and ensuring stable economy. It commits to reduce 50 percent of Pakistan's overall projected GHG emissions by 2030 with 15% from the country's own resources and 35% subject to provision of international grant finance that would require USD 101 billion just for energy transition The high priority actions focus on mitigation and adaptation, including measures such as renewable energy, electric vehicles, nature-based solutions, flood risk mitigation, and enhancing water recharge and enhancing protected areas The NDC recognizes the climatic impact drivers in the plains of Punjab (among other provinces) with extended and frequent riverine floods, heatwaves relevant to agriculture and health, coupled with increased aridity in arid and semi-arid regions. It underlines the degrading ecosystems affecting human health, adversely impacted water-agriculture, and the reduced productivity of ecosystems It highlights the elements, including the ones highlighted in the Rajanpur case study, are important to understand climate vulnerability, threats, and needs. It mentions that extreme events are becoming more frequent, incurring immediate as well as long-term L&D The L&D toolkit can help calculate the economic as well as the non-economic cost of L&D that is not catered for as part of the NDC

(continued)

Table 1 (continued)

Climate change policies and strategies	*Analysis of loss and damage components*
Framework for the Implementation of the Climate Change Policy (2014–2030)	The framework documents the vulnerabilities of various sectors and identifies appropriate actions under adaptation and mitigation From the six pillars identified in the Rajanpur case study, the framework mentions agriculture, health, livestock, and migration. It mentions infrastructure from the point of view of building climate resilience in the water sector (reservoirs, irrigation, rainwater harvesting, etc.) and critical infrastructure like dams, barrages, irrigation network, roads, bridges, railway lines, power stations, and river embankments. Also, coastal and urban infrastructure has been mentioned While specific actions have been proposed to address vulnerabilities under various pillars but specific measures to address L&D need to be included in the framework for implementation
Draft Punjab Climate Change Policy (2017)	The draft policy touches on some of the pillars identified to build resilience. However, an L&D argument is not articulated as part of the policy. Recent developments in COP 27 have paved way for provinces like Punjab that are vulnerable to extreme events including floods and heatwaves, to weave in L&D as part of the policy discourse

Climate change policies and strategies	*Analysis of loss and damage components*
Draft Punjab Climate Change Action Plan (2021)	The draft plan does have four of the six elements highlighted in the Rajanpur case study, but specific reference to rural infrastructure (especially housing) and micro-enterprises have not been given much focus It talks about loss of biodiversity, infrastructure, irrigation, energy transmission/distribution, and livelihood, but the mention and integration of the concept related to L&D remains absent
Draft National Adaptation Plan (NAP)	It will be a mechanism to cope with climate challenge and to reduce vulnerabilities with a comprehensive medium and long-term plans. It is under development and is expected to be finalized by June 2023. The findings from the case study can be extremely useful to build a narrative around L&D and to have measures in place focusing on the six pillars to have measures in place to avert, minimize, and address L&D

Damage associated with Climate Change Impacts (WIM)[13], focusing on enhancing knowledge and understanding of comprehensive risk management approaches, fostering dialogue, coordination, coherence, and synergies among all relevant stakeholders, and enhancing action and support including finance, technology, and capacity building. Policy to address Loss and Damage due to extreme events such as floods can be classified into measures that (a) avoid damages due to floods, (b) minimize the destruction, and (c) policy to cater to residual impacts that cannot be avoided.

The objective of the Santiago Network (SNLD)[14] is to catalyse the technical assistance of relevant organizations, bodies, networks, and experts, for the implementation of relevant approaches for averting, minimizing, and addressing Loss and Damage at the local, national, and regional levels. There is a need to mainstream these in the context of Loss and Damage in national policy domain. With the Loss and Damage fund established at COP27, there is a need to revise the policy to have a section focusing specifically on Loss and Damage, with the recent floods in 2022 as a case highlighting the limits to adaption. There is a need to be explicit when referring to the impacts of climate change that are un-avoided or unavoidable.

The six pillars identified through the implementation of the toolkit in Rajanpur can form the basis to recognize Loss and Damage, as distinct from adaptation. The policy emphasises the need to mobilise action and support to address Loss and Damage focusing on the elements of resilience including agriculture, livestock, human health, infrastructure (especially rural housing), migration, and micro-enterprises. Many of these elements are interlinked; impact on one can have an amplifying effect on the other. Hence, a synergistic approach is needed, where these pillars are considered part of one system and must be taken forward simultaneously.

From the review of the case study from the perspective of the six pillars, the impact of floods on women, the elderly, children, and the marginalised is much more intense. Each of the elements further enhances their vulnerability—with houses destroyed, they are living under shelter in

[13] https://unfccc.int/topics/adaptation-and-resilience/workstreams/loss-and-damage/warsaw-international-mechanism.

[14] https://unfccc.int/santiago-network.

poor conditions and often dealing with health impacts (including mental health). Pregnant and lactating mothers are at risk of reduced nutritional intake, leading to malnutrition in newborns. Loss of livestock and agricultural land lead to income losses and nutrition intake for their families. Many engaged in household enterprises, such as embroidery, don't have the market to sell their products. Migration in many cases, leads to social constructs for females being challenged with a feeling of helplessness and deprivation.

The case study highlighted that repeated floods diminished people's ability to sustain their livelihoods. While some households sustained larger absolute impacts, poeple experiencing poverty were impacted severely from a relative point of view. While some could undertake preventive measures, others never expected the next floods to be even more extreme. The availability of financing mechanisms, including limited to no insurance instruments, was a major impediments for many. Some received support from local organizations but the community highlighted very limited financial and technical support from the government. The role of the private sector in ex-ante disaster risk insurance needs to be weaved as one of the most essential policy options to cope with disasters, help in recovery, and build resilience.

Knowledge and information gathered through the Rajanpur case study and many more cases from developing countries facing climate catastrophesshould be synthesised for wider dissemination. These can serve as building blocks for the Transition Committee on Loss and Damage work to help define climate-induced Loss and Damage and means to address it.

The toolkit used to assess Loss and Damage in Rajanpur and many other tools developed by other organizations could be reviewed and a standardized toolkit could be developed to calculate the economic and non-economic Loss and Damage. Such an instrument could help address the financing question of 'how much' and 'to whom'. Uniformity in approaches for assessment will need to be designed and agreed upon.

Acknowledgements The manuscript builds upon the earlier work through the APN-supported project to develop a *Methods Toolbox to Assess Loss and Damage*, and for its application in India, Nepal, and Pakistan. It takes the learning from the Pakistan case study and links to the current deliberations on Loss and Damage fund. I would like to thank Ali Tauqeer Sheikh (Member, Pakistan

Climate Change Council) for his valuable input and guidance in developing this manuscript.

The APN-supported project was led by LEAD Pakistan, in partnership with AIDMI-India, IDS-Nepal, and UNU-EHS Germany. I would like to acknowledge contributions from team members including Ali Tauqeer Sheikh, Anam Zeb, and Arif Rehman of LEAD Pakistan, Mihir Bhatt and Vishal Pathak of AIDMI, Dinesh Devkota and Prakash Koirala of IDS-Nepal, and Kees van der Geest, Dr. Koko Warner, Prof. Dr Jakob Rhyner, and Markus Schindler of UNU-EHS Germany for the execution of the project.

References

Grantham Research Institute on Climate Change and the Environment (2022). What is climate change 'Loss and Damage'? The London School of Economics and Political Science.

Nadeem, O, Jamshed, A., Hameed, R., Anjum, G.A., Khan, M.A (2014). Post-flood rehabilitation of affected communities by NGOs in Punjab, Pakistan-learning lessons for future. *J Facul Eng Tech* 21 (1):1–20.

UNFCCC (2012). Decision 3/CP.18: Approaches to address loss and damage associated with climate change impacts in developing countries that are particularly vulnerable to the adverse effects of climate change to enhance adaptive capacity.

World Meteorological Organization (2021). The Atlas of Mortality and Economic Losses from Weather, Climate, and Water Extremes (1970–2019).

CHAPTER 5

A Conceptual Framework and Research Design for Assessing Losses and Damages from Climate Change in Vulnerable Communities

Kees van der Geest

Abstract The purpose of this paper is to present a conceptual framework and people-centred research design for assessing climate-induced losses and damages in vulnerable communities. The mixed-method research design builds on progressive insights from 12 fieldwork-based, empirical case studies, conducted across the Global South, in which more than 4000 respondents participated. The paper discusses key concepts, such as sudden-onset events, slow-onset processes, vulnerability, disaster risk reduction, coping strategies, adaptation and adaptation limits and constraints, and brings these together in a conceptual framework for assessing losses and damages in vulnerable communities. Subsequently,

K. van der Geest (✉)
United Nations University Institute for Environment and Human Security, Bonn, Germany
e-mail: geest@ehs.unu.edu

V. Wijenayake et al. (eds.), *Linking Climate Change Adaptation, Disaster Risk Reduction, and Loss & Damage*,
https://doi.org/10.1007/978-981-99-8055-0_5

the paper discusses research design considerations, such as purposive study site selection and random sampling of survey respondents. Lastly, it presents hands-on research tools, which include a household questionnaire, focus group discussions, expert interviews, participatory evaluation of climate action and open interviews to elicit personal testimonies of how climate change induces losses and damages and how this affects people's lives and livelihoods.

1 Introduction

'Loss and Damage' is an emerging topic in climate change negotiations, research and policy and the implementation of climate action. The world has entered in the third era of climate change, which is the era of Loss and Damage. In the first era, the era of mitigation, the hope was that we could avoid dangerous climate change by reducing greenhouse gas emissions. In the second era, the era of adaptation, we thought we could adapt to the climatic changes that were not avoided through mitigation. In the third era, we have realized that in some parts of the world, adaptation is increasingly difficult and sometimes impossible, and losses and damages have become inevitable there. In line with this, Loss and Damage[1] is now considered—alongside mitigation and adaptation—a third pillar of the United Nations Framework Convention on Climate Change (UNFCCC). It is still of utmost importance to reduce greenhouse gas emission to limit global warming (averting losses and damages), and to adapt to climatic changes where possible (reducing losses and damages). But we now also have to find ways to deal with unavoided and unavoidable losses and damages, and to protect the most vulnerable people (addressing losses and damages). There is an important climate justice element to this, as the people and countries that have least contributed to global warming are often the most likely to experience losses and damages from climate change (Boyd et al., 2021).

[1] In this paper, we follow the IPCC AR6 distinction between Loss and Damage (L&D) and losses and damages (l&d). L&D—singular and capital letters—refers to "political debate under the UNFCCC following the establishment of the Warsaw Mechanism on Loss and Damage in 2013"; l&d (plural and small letters) refers to "harm from (observed) impacts and (projected) risks and can be economic or noneconomic" (IPCC WGII AR6 Glossary).

At COP27 in 2022, a Loss and Damage Fund was established to support vulnerable countries that are most severely affected by climate change impacts. In the words of Simon Stiel, Executive Secretary of the UNFCCC, the fund is meant to "address the impacts on communities whose lives and livelihoods have been ruined by the very worst impacts of climate change" (UNFCCC, 2022). An important step towards meaningful action to address losses and damages in such vulnerable communities is to improve our ability to identify and assess these losses and damages.

As loss and damage is a relatively new area of research in the field of climate change (for an overview of the literature, see McNamara and Jackson, 2019), there are currently no well-developed assessment methods available to countries and organizations. This paper proposes a fit-for-purpose and adaptable research design for assessing losses and damages from climate-related hazards, with a focus on vulnerable communities.

The paper aims to advance methodologies for assessing losses and damages in the context of climate change, by building on the experiences of the first-ever multi-country assessment of losses and damages in vulnerable communities, which included case studies in The Gambia, Burkina Faso, Ethiopia, Kenya, Mozambique, Nepal, Bangladesh, Bhutan and the Federated States of Micronesia (Warner and van der Geest, 2013). Based on the experience of the first generation of case studies, a methods handbook was prepared and tested in Nepal, Pakistan and India (**van der Geest and Schindler, 2016, 2017). This paper builds on the cumulative learning-by-doing approach adopted in these case studies.

The paper argues that in order to effectively minimize and address losses and damages, we need to have a deeper understanding of the causes as well as the consequences. This involves examining the combination of physical stressors and social vulnerabilities that contribute to losses and damages, such as exposure to shocks and a lack of coping and adaptive capacity. It also involves a deep understanding of the values, perceptions, needs and preferences of people living at the frontlines of climate change. Studying these factors requires considering a range of disciplines beyond climate science. There is an urgent need to bring in research skills and experiences of anthropologists, psychologists, development scholars and other social scientists with a community-based approach to enhance our understanding of the consequences of losses and damages and set priorities for addressing losses and damages together with affected communities (Tschakert et al, 2019).

At this point, it is essential to ask ourselves the question: what do we actually mean by *assessing* losses and damages? According to the Oxford Learners Dictionary, the verb 'to assess' means: "To estimate the nature, quality or value of something" (Oxford Learners Dictionary, 5th edition). What I want to emphasize here is that to assess something is not the same as to *measure* something, which, according to the same dictionary, means "to find the size, length or amount of something by comparing to a standard unit" (Oxford Learner's Dictionary, 5th edition). A central premise of this paper is that a proper assessment of losses and damages involves measuring what one can measure and documenting and understanding what one cannot measure. I argue that understanding losses and damages, with the aim of averting, reducing and addressing it, is at least as important as measuring it. Importantly, there is a wide range of losses and damages that cannot even be measured adequately. If the focus of a loss and damage assessment would be purely on measuring losses and damages, all these 'non-economic losses and damages' (NELDs) would stay under the radar (Chandra et al., 2023).

To properly understand the losses and damages that individuals, households, communities, business and other actors incur, an assessment should focus not just on *what* is lost and damaged, but also on *how* and *why*. This shifts the purpose of the loss and damage assessment to a stronger consideration of the adaptation limits and constraints that vulnerable communities face (Berkhout and Dow, 2022; Mechler et al., 2020).

1.1 Limitations

The research design presented in this paper is suitable for assessing losses and damages in vulnerable communities, where severe climate impacts are already a present-day reality; where adaptation limits are approached or already crossed; and where significant constraints to adaptive capacity exist. So the focus is on people and losses and damages that are already occurring. For research on projected risks of future losses and damages, and for more quantitative and less people-centred assessments of losses and damages at higher levels of scale (national, global), other methods are needed (see e.g. Bouwer, 2019). Similarly, for research on loss and damage to ecosystems, other methods are needed (see Janzen et al., 2021).

This paper uses 'losses and damages' as a composite term, and does not attempt to distinguish losses from damages. Some scholars have explored the added value of separating the two, defining losses as permanent or irreversible and defining damages as repairable (Doelle and Seck, 2019; McNamara and Jackson, 2019; Tschakert et al., 2019; Puig, 2022). There may be value in doing this, but there are also important drawbacks, as there is a large gap between this conceptual distinction and the language reality. For example, we tend to speak of 'loss of livelihood' when people's livelihoods are hit by a drought, flood or storm. However, when we say this, we do not want to imply that these livelihoods are lost forever.

1.2 Outline

The structure of this paper is as follows. First, key concepts are discussed and brought together in a conceptual framework for assessing losses and damages in vulnerable communities. Second, the paper provides guidance on the overall design and organization of fieldwork activities, including site selection, sampling of survey respondents and training of research teams. Third, the different work streams and methods for data gathering are discussed, including desk study, the household questionnaire, focus group discussions, expert interviews, story-telling tools and a method for participatory evaluation of climate action by governments and NGOs.

2 Key Concepts

This section gives an overview of key terms used in relation to loss and damage from climate change. Many of the key concepts originate from other thematic areas in global change research, particularly disaster risk reduction and climate change adaptation. The work on loss and damage has the potential to link these fields, but to do so effectively, it is important to create conceptual clarity. After introducing and defining the key terms, they will be placed in a conceptual framework that informs the research design for loss and damage assessment proposed in this paper. The key concepts fall in different domains: climate, vulnerability, impacts and responses.

2.1 Climate

As this paper considers losses and damages as a result of actors' insufficient capacity to respond to different climate stressors, it is important to distinguish the different types of climatic stressors in the first place. Each of these stressors types have different time dimensions and require different responses. First, we have climate hazards and risks which are more or less permanent characteristics, and which require ex-ante risk management strategies by households and governments and organizations. Second, there are sudden-onset climatic events that occur at a specific point in time, and which require coping strategies at household level and emergency responses by governments and organizations. Thirdly, there are longer-term or slow-onset changes in climatic conditions that require adaptation by households, governments and organizations.

Hazard: The potential occurrence of an anthropogenic or non-anthropogenic physical event that may cause loss of life, injury or other health impacts, as well as damage and loss to property, infrastructure, livelihoods, service provision and environmental resources (IPCC, 2012).

Risk: The potential for consequences where something of human value (including humans themselves) is at stake and where the outcome is uncertain. Quantitatively, risk is often represented as probability of occurrence of a hazardous event(s) multiplied by the consequences if the event(s) occurs (IPCC, 2014).

Disaster: Severe alterations in the normal functioning of a community or a society due to hazardous physical events interacting with vulnerable social conditions lead to widespread adverse human, material, economic or environmental effects that require immediate emergency response to satisfy critical human needs and that may require external support for recovery (IPCC, 2012).

Sudden-Onset Events: These take place over short time frames (typically hours, days or weeks). Climate-related sudden-onset events include floods, cyclones, tornadoes, landslides, sudden collapses of riverbanks, extreme rainfall events, heat waves and wild fires.

Slow-Onset Processes or Changes: These take place over longer time frames (typically years to decades). Climate-related examples are sea level rise, salinization, ocean acidification, temperature rise, desertification and changing rainfall patterns. Droughts are usually categorized as slow-onset phenomena, but in terms of the consequences and the coping strategies adopted in response, a drought is often similar to sudden-onset events (van der Geest and van den Berg, 2021).

2.2 *Vulnerability*

To study and understand how climatic events result in life-threatening losses and damages for some, and hardly a scratch for others, the concept of vulnerability is crucial. The IPCC defines vulnerability as "the propensity or predisposition to be adversely affected. Vulnerability encompasses a variety of concepts and elements including sensitivity or susceptibility to harm and lack of capacity to cope and adapt" (IPCC, 2014: 1775). In one of the first papers on vulnerability to climate change, Robert Chambers (1989) distinguished external and internal vulnerability and this distinction is particularly relevant for assessing losses and damages in vulnerable communities. Chambers describe the external side of vulnerability as exposure and the internal side as a lack of coping capacity (Chambers, 1989). Current thinking and studies about vulnerability still use this distinction, and add a third element: sensitivity (Füssel and Klein, 2006).

Sensitivity refers to "the degree to which a system is affected, either adversely or beneficially, by climate-related stimuli. The effect may be direct (e.g , a change in crop yield in response to a change in the mean, range, or variability of temperature) or indirect (e.g., damages caused by an increase in the frequency of coastal flooding due to sea level rise)" (IPCC, 2007).

Similar to the distinction between the internal and external side of vulnerability, it is helpful to distinguish collective vulnerability and individual vulnerability: Collective vulnerability results from area-level variables that are the same for all households in a given community, region or country. By contrast, the level of individual or household vulnerability differs between households in a community (Adger, 1999; McLeman, 2010).

2.3 *Responses*

To study losses and damages from climate changes as impacts beyond or despite actors' efforts to avoid them, the research design presented in this paper distinguishes three broad categories of responses to different climatic stressors: preventive measures, coping measures and adaptation measures.[2]

Preventive Measures/Ex-ante Risk Management: This involves the actions households take to prevent or minimize future impacts of climate changes and extreme weather events. Classic risk management theory distinguishes four ways of dealing with risk that are usually adopted as a function of the probability and severity of events (Bekefi et al., 2008). Risk avoidance: Refrain from certain activities or avoid certain places or situations because the probability of severe impacts is too high. Risk reduction: Actions taken to lessen the probability and/or negative consequence associated with a risk. Risk transfer: Sharing the burden of a loss with others, including insurance solutions and community-based systems related to social capital. Risk retention: Accepting a risk because of low probability and non-severe impacts that make it possible for adverse effects to be absorbed.

Coping Strategies: Defined by the IPCC (2012) as "The use of available skills, resources, and opportunities to address, manage, and overcome adverse conditions, with the aim of achieving basic functioning of people, institutions, organizations, and systems in the short to medium term".[3] Literally, to cope means 'to deal successfully with something difficult' (Oxford Learner's Dictionary, 5th edition). In livelihood research and development studies, the term is reserved for the things people do in the aftermath of adverse events (such as a flood or drought-induced crop failure) to survive or 'get back to normal'. Examples are: selling assets, migration, reliance on alternative sources of food and income when the main source of livelihood fails (e.g. food aid, wild foods, petty trade, labour), taking loans, etc (Warner and van der Geest, 2013).

Adaptation: Changes in social-ecological systems in response to actual and expected impacts of climate change in the context of interacting non-climatic changes (Moser and Ekstrom, 2010). Several types of adaptation measures can be distinguished:

- Autonomous and Planned Adaptation: Planned adaptation involves actions and deliberate policy by public bodies (e.g. governments,

[2] The relationship between preventive strategies, coping and adaptation is described in detail in van der Geest (2004: 20–29).

[3] This glossary entry builds from the definition used in UNISDR (2009) and IPCC (2012a).

NGOs) to protect citizens against climate change and its impact (Smit et al., 2001). By contrast, autonomous adaptations are natural or spontaneous adjustments to climatic changes by individual actors, such as households, small enterprises or communities[4] (Carter et al., 1994, in Fankhauser et al., 1999: 69).

- Pro-active and Reactive Adaptation: Pro-active or anticipatory adaptation measures are adopted in response to future expected climate, and before impacts have occurred. Reactive adaptation measures are adopted in response to climatic changes or events that have already had adverse effects or caused damage (Füssel, 2007).

Many studies use the terms coping and adaption synonymously (Birkmann, 2011). This is problematic because they involve different types of responses to different types of stresses (van der Geest and Dietz, 2004). Coping strategies are short-term responses to the impacts of sudden or unusual events. By contrast, adaptation refers to longer-term adjustments to more permanent changes in the climate.[5]

Table 1 shows how the three types of responses (prevention, coping and adapting) connect to different types of climate stresses, and provides some examples of these responses in practice. There are multiple linkages between different types of household responses to climatic stressors. First, the success of ex-ante preventive measures determines the need for and success of ex-post coping strategies. Second, short-term coping measures can evolve into more permanent livelihood adaptations when they become recurrent. Third, when households change their preventive measures in response to changes in perceived risk, they are in fact adapting.[6]

[4] The term autonomous adaptation is also used for spontaneous adjustments in ecosystems (see e.g. IPCC WG2 AR5, Chapter 4).

[5] For the more elaborate definition of adaptation we used in the case studies, see Moser and Ekstrom (2010). Their definition recognizes that adaptation measures are often adopted in response to a mix of climatic and non-climatic changes and aim to meet more than climate goals alone.

[6] For an overview of linkages between prevention, coping and adapting, see van der Geest and Dietz (2004). The framework is inspired by the early work of Susana Davies (1996) on 'adaptable livelihoods' in Mali.

Table 1 Different climatic stressors require different household responses

Climatic stressor	*Household response*
Climate variability, risk and hazards • Normal climate/weather uncertainties • Normal risk of extreme weather events	**Preventive measures (disaster risk reduction)** • Physical protection, such as seawalls • Risk spreading in agriculture and livelihoods • Creating buffers • Investing in social safety nets
Climate-related events (disasters) • Floods • Droughts • Cyclones/storms • Landslides	**Coping strategies** • Selling assets • Alternative income sources • Reliance on social networks • Food aid and other relief
Climatic changes • Changes in 'average' conditions • Changes in risk (frequency and severity) of extreme weather events	**Adaptation** • Agricultural change • Livelihood diversification • Migration • Changes in 'normal' risk management, including • preventive measures and coping strategies

Source Author's own, building on van der Geest and Dietz (2004)

2.4 *Impacts and Loss and Damage*

Though there is much overlap between impacts of climate change and loss and damage from climate change, the two terms are not the same (Roberts et al., 2014; Zommers et al., 2016). The concept of loss and damage emphasizes that avoidable impacts have not been avoided and that some impacts cannot be avoided even with large improvements in mitigation and adaptation policy.

The IPCC defines impacts as "effects on natural and human systems of physical events, of disasters, and of climate change… Impacts generally refer to effects on lives, livelihoods, health, ecosystems, economies, societies, cultures, services, and infrastructure due to the interaction of climate changes or hazardous climate events occurring within a specific time period and the vulnerability of an exposed society or system" (IPCC WG2 AR5 glossary).

The IPCC definition of losses and damages is, in my view, problematic because it does not adequately distinguish the concept of 'loss and damages' from 'impacts'. According to the IPCC, losses and damages "refer broadly to harm from (observed) impacts and (projected) risks and can be economic or noneconomic" (IPCC, 2022). To set the concept aside from 'impacts', it is necessary to emphasize the residual aspect of loss and damage, by emphasizing that losses and damages incur despite efforts to avoid them. A concise definition could be that losses and damages refer to "adverse effects of climatic stressors that occur despite mitigation and adaptation". A more precise definition would be that *losses and damages refer to adverse effects of climatic stressors resulting from inadequate efforts to reduce greenhouse gas emissions and insufficient capacity to reduce the risks associated with climatic stressors, to cope with impacts of climatic events and to adapt to climatic changes.*

Several types of loss and damage can be distinguished, for example:

Economic and Non-economic Losses and Damages: The main distinction between the two categories is whether the lost items are commonly traded in markets (Fankhauser et al., 2014). Table 2 gives an overview of different types of non-economic losses and damages.

Avoided, Unavoided and Unavoidable Loss and Damage: According to Verheyen (2012), there are three types of losses and damages: avoided, unavoided and unavoidable. The term 'avoided losses and damages' refers to impacts of climate change that have been avoided by mitigation and adaptation action. Unavoided losses and damages could

Table 2 A categorization of different types of non-economic losses and damages (NELD)

Human life	*Meaningful places*	*cultural heritage*	*Intrinsic values*	*Biodiversity*	*Ecosystem services*
Lives	Territory	Heritage	Dignity	Species	Landscapes
Health	A place to call 'home'	Traditions	Agency	Biodiversity	Productive land
Well-being	Sacred sites	Customs	Identity	Habitats	Supporting services
Livelihoods		Culture	Social cohesion	Flora	Regulating services
		Indigenous knowledge	Security	Fauna	Provisioning services
			Sovereignty		Cultural services

Source The table builds on past and ongoing work at UNU-EHS, with inputs from Magdalena Mirwald and Cathleen Eberle

have been avoided, but have not been avoided because of inadequate mitigation and adaptation efforts. Lastly, there are losses and damages that are unavoidable no matter how ambitious mitigation and adaptation efforts are. Avoidable losses and damages can be addressed by more effective adaptation measures, removing adaptation barriers, improved disaster risk reduction and increasing resilience and coping capacity of vulnerable communities. However, those impacts that are either unavoided or unavoidable will need to be addressed by a range of other approaches, such as social protection, planned relocation, assisted migration, insurance solutions and compensation. The establishment of the Loss and Damage Fund is an important step towards operationalizing these approaches.

Some key terms in the emerging literature on loss and damage are closely related to adaptation such as:

- Maladaptation: "Action taken ostensibly to avoid or reduce vulnerability to climate change that impacts adversely on, or increases the vulnerability of other systems, sectors or social groups" (Barnett and O'Neill, 2010: 211).
- Erosive Coping: Coping strategies are erosive when they undermine future livelihood security (van der Geest and Dietz, 2004; Opondo, 2013). Examples of this include selling productive assets to buy food and taking a child out of school to beg or work in the informal sector.
- Adaptation Constraints: Factors that make it harder to plan and implement adaptation actions or that restrict options (IPCC WG2 AR5 glossary).
- Adaptation Limit: The point at which an actor's objectives (or system needs) cannot be secured from intolerable risks through adaptive actions (IPCC). This is the case when "no option exists, or [when] an unacceptable measure of adaptive effort is required" (IPCC, 2014: Ch. 16, p. 8). Soft Limit: No adaptive actions are possible to avoid intolerable risks. Hard Limit: Options are currently not available to avoid intolerable risks through adaptive action (IPCC WG2 AR5 glossary).

3 Conceptual Framework

The framework discussed in this section connects losses and damages from climate-related stressors to vulnerability, disaster risk reduction, coping strategies and adaptation (Fig. 1). The framework results from progressive insights from working on loss and damage in vulnerable communities in the past ten years (Warner and van der Geest, 2013; van der Geest and Schindler, 2017; van der Geest and Warner, 2015, 2020), and previous work on impacts of and adaptation to climate change in dryland West Africa (van der Geest 2004, 2011; van der Geest and Dietz, 2004). This work, in turn, builds on a longer tradition of studying livelihoods in risk-prone environments that emerged in the 1990s (Chambers, 1989; Davies, 1996; Blaikie et al., 1994; Scoones, 1998; Ellis, 1998).

The blue box in the upper part of Fig. 1 shows the vulnerability context of households and communities that shape households' livelihood strategies and the measures they put in place to reduce the risk of being adversely affected by climatic and other stressors. The framework distinguishes collective vulnerability—resulting from area-level variables that are the same for all households in a given community—and individual or household vulnerability (Adger, 1999; McLeman, 2010). When a region experiences adverse slow-onset processes or when sudden-onset extreme weather events hit, some households will experience impacts (such as a crop failure or damage to properties) while others may not. This depends on their vulnerability profile—particularly their exposure—and the measures the household has adopted to reduce risk prior to the event. When the household experiences no impact, there is also no loss and damage (hence the green colour of the 'no-impact-box'). When the household is affected by the climatic stressor, it may incur or avoid residual loss and damage depending on whether effective measures are adopted to adjust (hence the red-green colour of the impact and coping/adaptation boxes).

In the case of sudden-onset events, household responses to deal with impacts are 'coping strategies'. If there are no impacts, there is also no need to cope. This is different in the case of slow-onset changes. Households can adapt in response to *actual* or *expected* impacts (Moser and Ekstrom, 2010: 22026). Reactive adaptation is in response to actual impacts and pro-active adaptation is in response to expected impacts.

If there is nothing the household can do to cope or adapt, it will incur losses and damages (hence the red colour of the no-adaptation

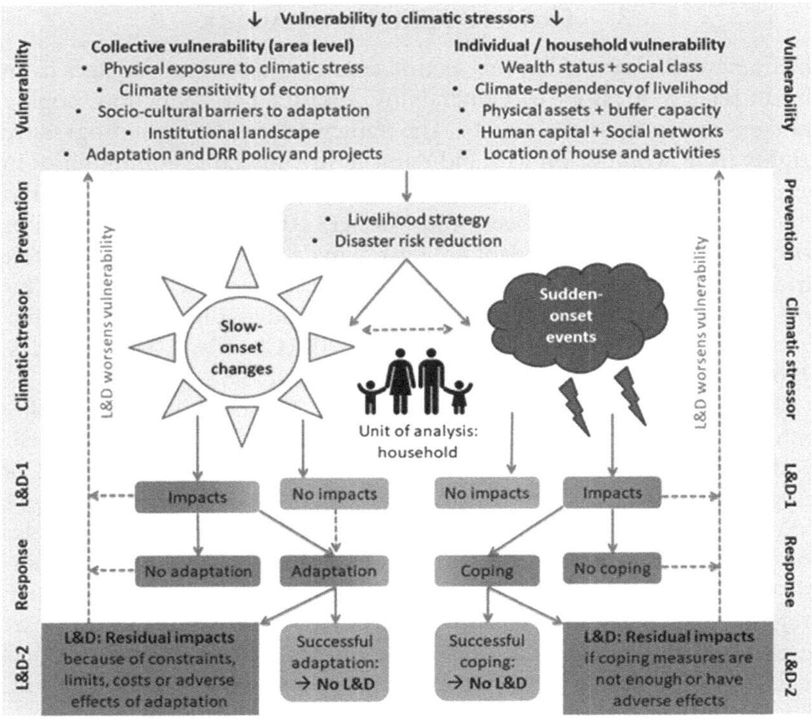

Fig. 1 Conceptual framework

and no-coping boxes). If coping or adaptation measures are adopted, these may or may not be effective in avoiding residual loss and damage, depending on the household's adaptive capacity and the magnitude of the climatic stressor (or in other words: adaptation constraints and limits[7]). If measures are insufficient, costly or 'erosive' in the longer term, households incur loss and damage (Warner and van der Geest, 2013). Lastly, there is a feedback loop connecting loss and damage back to the

[7] IPCC's Fifth Assessment Report defines adaptation constraints as "factors that make it harder to plan and implement adaptation actions" and it defines an adaptation limit as "the point at which an actor's objectives ... cannot be secure from intolerable risks through adaptive actions" (See also Dow et al., 2013).

household's vulnerability profile. This is because the losses and damages incurred render the household more vulnerable in the face of ongoing climatic changes and future extreme events.

A key insight from Fig. 1 is that households can incur loss and damage in two ways. When households incur impacts of climatic stressors despite the risk reduction measures and pro-active adaptation measures they adopted to avoid such impacts, we speak of first-order losses and damages (LD-1). The second way in which households incur losses and damages is when the coping strategies and adaptive measures they adopt have costs or negative side-effects that affect livelihood sustainability in the longer term (erosive coping and maladaptation). These are called second-order losses and damages (LD-2). Alternative one can speak of direct and indirect losses and damages.

4 Research Design

This section provides guidance on how to design a mixed-method, people-centred research project that documents already occurring losses and damages from climate change in vulnerable communities. The actual research tools are discussed in the subsequent section.

4.1 Scale

This research design in this paper has been prepared for place-based assessments of losses and damages in local, community-based case studies. However, with sufficient human and financial resources and smart sampling techniques, there is potential for scaling up to regional and even national level. The case studies that informed this paper were generally conducted in one district (or similar administrative unit) per country. In each district, a select number of villages were surveyed. The budget for fieldwork was usually between 20,000 and 35,000 euro per test country. With more budget, a larger study area can be covered, or more households within the study area can be surveyed. Research teams typically consisted of a principal investigator, four to eight enumerators, a note taker for the qualitative interviews and a logistics manager. The average duration of the fieldwork was three to four weeks per case study. The teams conducted approximately 150 to 350 household questionnaires, five to ten focus group discussions (FGDs) and five to twelve expert interviews (EIs).

4.2 Research Domains

As explained in the introduction, to properly assess losses and damages in vulnerable communities, it is not enough to simply document items that are lost or damaged. Data need to be gathered and analysed in seven different research domains. For each domain, ideally, data need to be gathered and analysed at household level, but also at district or regional level. The type of data needed in each domain is shown in Table 3.

The household questionnaire covers all seven domains at household level, and generates a combination of quantitative data and more qualitative information from open questions. In-depth insights on each domain at the level of individual people can be obtained through 'life history' type of open interviews, which we have also called 'livelihood histories' (van der Geest, 2004; Ayeb-Karlsson et al., 2016; Ahmed et al., 2019). Additional research tools, such as desk study, focus group discussions and expert interviews can be used to gather and analyse information at the district or regional level, including the effectiveness of planned adaptation and emergency support by the government and organizations.

4.3 Site Selection

As this paper aims at improving methods for enhancing our understanding of what losses and damages look like on the ground, site selection is crucial. This section lists a set of criteria for selection of case study areas. In empirical studies of loss and damage, the site selection should be 'purposive' (not random), while selection of survey respondents within study sites should be random. This way, the result can stand academic tests and at the same time provide relevant insights.

Relevance: The study area must have experienced and be vulnerable to climate hazards. The research team has the option to study the losses and damages associated with a certain event (e.g. a cyclone or flood that hit the area in a certain year) or focus on a broader range of climate-related stressors.

Data Availability: Availability of at least 30 years of meteorological and/or hydrological data is highly desirable. Depending on the focus of the study, daily data on rainfall, temperature, wind speed, water levels and/or river flow volumes would be used to provide a more solid scientific base for linking losses and damages to climatic stressors. It is extremely helpful to conduct a pre-fieldwork analysis of meteorological and/or

Table 3 Research domains

Research domain	*Questionnaire and in-depth interviews*	Example	*Desk study, focus groups and expert interviews*	Example
Climatic stressors, including sudden-onset events and slow-onset processes	People's perceptions	*Changes in rainfall patterns*	Meteorological data	*Drought risks, flood extent; sea level rise*
Vulnerability of livelihoods to impacts of these stressors	Household vulnerability	*Distance between house and eroding river bank*	Area-level data	*Data from ministries on crop production in drought years*
Preventive measures to deal with existing climatic risks and variability	Household measures	*Mixed cropping; protecting houses*	Information on planned measures by organizations	*Early warning systems*
Loss and damage from direct impacts of climatic stressors despite preventive measures	At household level	*Harvest failure, loss of properties; damage to houses*	At area level	*Loss of lives, damage to infrastructure, number of displaced people*
Adaptation to climatic changes and impacts	By households	*Livelihood diversification*	By organizations	*Construction of sea walls*
Coping with impacts of climate-related events	By households	*Sale of livestock to buy food; migration*	By organizations	*Emergency relief or food aid*
Loss and damage related to the costs and adverse side-effects of adopted measures	To households	*Erosive coping measures*	At area level	*Maladaptation; adverse effects of emergency relief*

Source Author's own

hydrological data to ensure that the selected study area has, in fact, experienced climatic disturbances or deteriorations. For example, if the assessment focuses on loss and damage from changing monsoon patterns or increasing temperatures, it is important to ascertain, in advance of the fieldwork, that the meteorological data for rainfall and temperature indeed show adverse changes. If the research team does not have access to local meteorological data at the time of site selection, it is worth checking online databases with high-resolution environmental data (see e.g. de Sherbinin, 2014).

Climate Sensitivity: The population in the study area should have livelihoods that are sensitive to climatic perturbations. This is more typical in rural areas than in urban areas. The most climate-sensitive livelihoods occur in areas with predominantly rain-fed agriculture, in areas that are relatively remote with few opportunities for earning a non-farm income and in areas that are in unstable transition zones between agro-ecological systems (e.g. the desert margins and areas close to the permafrost limit).

Local Contacts: It is helpful if the implementing organization has well-established contacts in the study area, preferably through a local NGO or government agency with an office in the area or very nearby. Inhabitants and local leaders can also serve as valuable local contacts. While they may lack the formal and bureaucratic capacities, as well as the network of contacts that NGOs or government may have, they may provide unique insights to specific areas that NGOs or the government are unaware of. These contacts make it easier to organize the fieldwork and disseminate findings or organize follow-up activities afterwards.

Communication: The best research results are attained if at least some members of the team speak the same language as the study population. This should be considered when selecting a study area. Or vice-versa if a loss and damage assessment needs to be done in a particular place, it is important to recruit research team members who speak the local language.

4.4 Survey Sample

To obtain representative results from a selection (sample) of households in a certain location or area that can be generalized to the whole population, a systematic random sampling procedure is required. Random sampling means that each household in the study site has the same chance of being selected for an interview. There are different techniques for achieving this, for example:

- All households living in a location or area are listed, and a random selection is drawn from this list. This method can be convenient if a reliable and up-to-date list of households can be obtained from local authorities, other organizations or the population census. In the absence of such a list, when all households need to be identified by the research team, this method can be time-consuming.
- Households are selected from a map that is detailed enough to identify houses. All houses on the map are numbered and listed and a random sample is drawn from the list. For most locations in the world, the 'earth view' in Google maps can be used to create such a detailed map, but there may be quite a bit of 'noise' because not every physical structure is, in fact, a house (e.g. barns, livestock shelter, haystack, companies, public buildings, etc.). Moreover, a house can contain several households. A systematic approach is required to deal with this.
- Line Sampling: Lines are drawn through a location, and households are selected randomly along those lines, e.g. every fifth house on the left and right within 50 metres from the line. Some knowledge of spatial distribution of households (in terms of ethnicity, wealth, occupational groups, etc.) is required to draw the sample lines in a correct way.

In many cases, it will be desirable to sample households in two steps. Typically, the study area for a loss and damage assessment is not a single location, but a district, a municipality, a catchment area, an island or another area unit containing several localities. In this situation, a two-stage sampling procedure is desirable. The following example aims to illustrate this. If the area has a population of approximately 100,000 inhabitants in 50 localities, assuming an average household size of five (can be checked with census data), and a 1 per cent sample is required (i.e. 200 households), the team could decide to first select five localities and then randomly select 40 households per locality. The selection of localities can be random, but in many cases, it can be desirable to use a purposive selection procedure to make sure that certain characteristics that can influence vulnerability to climatic stressors (such as altitude, distance to main roads, distance to river, etc.) are well represented in the overall sample.

The choice of sampling methods mostly depends on the information that the research team has at its disposal and the time and human resources it is willing to invest in a quality random sample.

4.5 *Pilot Study*

It could be useful to split the fieldwork for assessing loss and damage into two parts: a 'pilot study', followed by the main research. During the 'pilot study', a small team, including the principal investigator, tests the questionnaire, prepares the sample framework, conducts one or two FGDs and organizes logistics for the arrival of the whole research team. Subsequently, possible lessons learned are considered, before the main research begins.

A standard questionnaire for assessing losses and damages from climate change, that should be relevant and applicable in a wide range of study sites, is publicly available. However, quite a lot of questions (e.g. on livelihoods, impacts and adaptation) are location- and climate-specific. The pilot study can be used to customize the questionnaire and adapt it to the local context and research needs.

4.6 *Training of the Fieldwork Team*

It is important that the enumerators and research assistants receive a proper training before they start interviewing households. Depending on their level of experience, typically, two to three days of intensive training should be enough to provide a good knowledge base for the fieldwork. The principal investigator should walk them through the methodology and particularly the questionnaire in detail. The team members should then discuss among themselves how to exactly ask questions in the field, and test the results of their work on each other to detect potential problems. When the research is conducted in areas with different languages, some extra time may be needed to ascertain if all questions can adequately be translated in the local languages.

5 Research Tools

This section describes the research tools used to assess losses and damages. It includes methods for quantifying losses and damages that are measurable and qualifying losses and damages that are not measurable. The assessment uses a mixed methods approach (Burke Johnson et al., 2007; Morgan, 2007). Six broad work streams are distinguished:

- Desk study
- Household questionnaire
- Participatory research tools, including focus group discussions
- Expert interviews
- Stories of Loss and Damage
- Participatory evaluation of planned climate action

5.1 Desk Study

The desk study consists of a literature review and an analysis of existing climate data (e.g. rainfall data) and, if available, impact variables (e.g. crop yields). The literature review should focus on relevant existing knowledge about impacts of climate change, risk management, coping mechanisms and adaptation in the study area and similar places nearby. In most cases, the climate threats we are studying are not new. Farmers in the Sahel, for example, have had to cope with drought impacts since time immemorial. Impacts of and responses to drought in the Sahel have been studied extensively, and assessments of losses and damages should build on such knowledge. Importantly, the assessment should go beyond this by also exploring limits and constraints to adaptation and the impacts of climate stressors that actors are not able to deal with. The literature review should give a brief overview of existing knowledge on this and identify knowledge gaps.

5.2 Household Questionnaire

The household questionnaire is a central tool in the research design for assessing loss and damage in vulnerable communities. The template questionnaire used in the 12 case studies that this paper builds on is about 10 pages long and the questionnaire interviews took approximately 45 to 60 minutes. The design and structure of the questionnaire follows the conceptual framework in Fig. 1. A balance is sought between measuring and understanding losses and damages, by gathering quantitative as well as qualitative information through open-ended and closed questions.[8] For example, the questionnaire first asks the respondent how a certain climate-related event, such as a drought, affected his or her household. This is an open question. After this, the questionnaire inquires in more detail about specific impacts (with pre-determined categories) and tries to quantify losses and damages if possible.

The questionnaire has been designed as a template that should be applicable across different types of study sites in rural areas of developing countries. Most questions in the questionnaire have relevance in all rural areas in developing countries. However, there are two sections in the questionnaire that need to be customized per study area. First, the closed questions about risk reduction and adaptation, and second, the closed questions about impacts. These sections need to be stressor-specific. For example, adaptation to sea level rise requires very different action than adaptation to increased drought risks. Similarly, impacts of a flood are quite different from impacts of a drought. Future users of this template questionnaire will need to phrase specific questions about impacts and adaptation for the climatic events and changes that their studies focus on (e.g. drought, flood, cyclone, sea level rise, heat waves).

Part 1: Household Info, Livelihood and Vulnerability: The questionnaire begins with the most basic socio-demographic information about the household, such gender, age, household composition, education level, marital status, etc. (max 1 page). This is important info as it helps to identify—in the analysis phase—which groups of people are most likely to suffer harm from climate-related stressors, and how this can inform policy to address losses and damages. The questionnaire continues with questions about people's sources of livelihood and more information that sheds a light on their exposure and sensitivity to climate-related hazards, such as the location and quality of their house. Data from this part of the questionnaire can be used as input for a multidimensional vulnerability index (MDVI) and related to the effectiveness of household measures to avoid and minimize losses and damages (see e.g. van der Geest and warner, 2015).

[8] Caution: The team leader has to be on top of the enumerators to ensure that the answers to the open questions are documented in sufficient detail.

Part 2: Climatic Stressors, Impacts, Responses and Loss and Damage: In this part of the questionnaire, we try to quantify what is quantifiable (crop losses, damage to houses and properties) and to qualify losses and damages that are not measurable (e.g. non-economic losses and damages (NELDs), such as loss of social cohesion, identity and mental health). This section includes questions about the following topics:

- Changes in the frequency and intensity of climatic stressors;
- Preventive measures to deal with 'normal' climate risks in the area and the effectiveness of these measures;
- Impacts of a specific climatic event on households' livelihood and well-being, e.g. crops, livestock, fishing, non-farm income, food prices, housing, properties, health;
- Coping strategies that households adopt to deal with unavoided impacts of climate-related events, and the effectiveness of these strategies (Fig. 2 provides an example of research results on this);
- Longer-term adaptation to climatic changes and impacts and their effectiveness; and
- Constraints to the effectiveness of preventive, coping and adaptation measures.

The questions about preventive measures, coping and adaptation were asked at two levels. First, we asked about households' own measures. After that, we inquired about the measures developed by organizations in the area.

As an example of research results from the questionnaire survey, Fig. 2 compares the ability of households to avoid losses and damages from droughts and floods in Kenya and the Gambia. The comparison between people who adopted certain coping strategies and those who didn't. An [*] indicates a statistically significant difference at $p < 0.05$ level. People who coped with droughts or floods by relying on non-farm income sources, for example, were less more successful at avoiding losses and damages. A policy implication could be that livelihood diversification can increase resilience to droughts and floods.

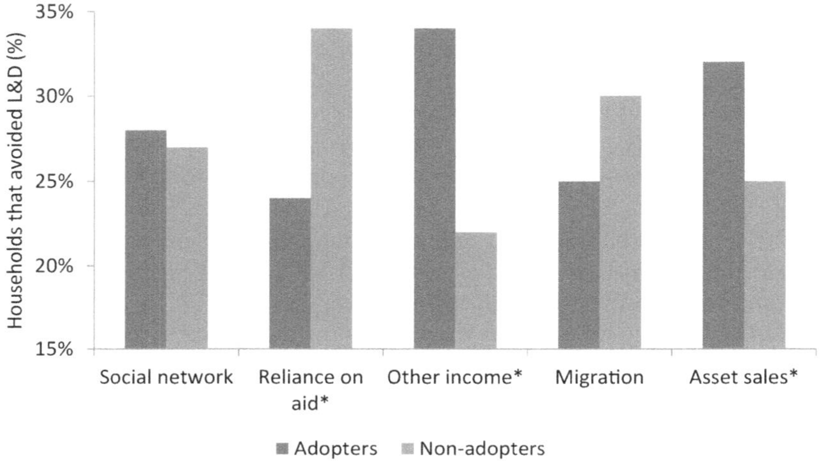

Fig. 2 Example of research finding about coping strategies

5.3 Participatory Research Tools

Participatory Research Approach (PRA) methods, such as focus group discussions, can be used to ask participants open questions that have more detailed, complex and sometimes contested answers. The insights from these research tools help the researchers better understand the dynamics between key concepts of this research (climate threats, impact, vulnerability, coping, adaptation and loss and damage). PRA tools complement the household survey (questionnaire) in that they yield more qualitative information on how climate variability and climate change can lead to losses and damages among local populations. Another advantage of conducting PRA sessions is that they depict the different experiences of men and women, young and old and possibly of different occupational groups (e.g. crop cultivators, pastoralists, labourers, traders) and socio-economic classes. This can be achieved by having separate groups for different categories of participants. Besides focus group discussions, there is a wealth of other PRA tools that can be useful for loss and damage assessments (see e.g. Rademarcher-Schulz et al., 2012; Afifi et al., 2016).

5.4 *Expert Interviews*

Expert interviews (EIs) should be conducted to obtain information that would not easily be obtained from PRA sessions and the questionnaire survey, particularly about planned adaptation measures and other interventions by governmental and non-governmental organizations that aim to reduce the risk of loss and damage or provide emergency support when disasters strike. Questionnaire respondents and participants in FGDs may be able to list some projects and interventions by government agencies and NGOs that aim to address adverse effects of climate variability and climate change in the study area. However, certain information about these interventions are often unknown to them. Expert interviews, including online, can also be useful in the pilot study phase when the main fieldwork and research instruments are prepared, and when information about climatic hazards, local livelihoods and typical adaptation measures is needed.

5.5 *Stories of Loss and Damage*

A select number of households could be interviewed in more depth to extract compelling stories of loss and damage. An efficient way to do this is to instruct questionnaire enumerators to alert the principal investigator when they conduct the questionnaire interview with a person who seems to have relevant experiences to share, for example, someone who has experienced specific adverse effects of climate change or someone who has successfully avoided losses and damages by adopting certain risk reduction or coping measures. These stories can be used in case study reports and other publications in the form of boxes and quotes.

Based on open interviews with checklists, the personal stories of loss and damage aim to give a face to the more quantitative data we gather, and that are presented in aggregates. The stories could take a life history or 'livelihood history' perspective. This method requires quite advanced interviewing and writing skills (a social science background is preferable), and can be time-consuming. Typically, a senior researcher or the principal investigator conducts these interviews.

5.6 *Participatory Evaluation of Climate Action*

The objective of this work stream is to assess the ability, effectiveness and constraints of existing climate change-related interventions by governments and NGOs in the study areas to avert, minimize and address losses and damages. This work stream complements the household questionnaire, which focuses more on autonomous adaptation and risk reduction measures by households themselves. Moreover, the exercise can inform possible solutions/actions for each research site, and local practitioners can learn a lot from the analysis as it sheds light on what works, what does not and why, from the point of view of intended beneficiaries.

This work stream builds on prior experiences from the Participatory Assessment of Development project (Dietz, 2013**; Pouw et al., 2016) and the Gibika Research to Action project (Ayeb-Karlsson et al., 2016). Data are gathered during FGDs and EIs. The work stream consists of four steps, in chronological order:

- A list of project or interventions is compiled during FGDs.
- Expert interviews are conducted to clean and complete the list.
- During FGDs, the effectiveness of these projects is evaluated from a people-centred perspective.
- This is followed by a needs assessment (part of the same FGD) to elicit people's views on what is needed to effectively avoid, minimize or address losses and damages in their communities.

This work stream is described in more detail in van der Geest and Schindler (2017).

6 Concluding Remarks

While the academic literature on losses and damages from climate change is growing fast, most of these studies use existing databases of disaster losses, often at national level. This type of research is useful in itself, but does not do justice to the perceptions, needs and preferences of the people who are most severely affected by climate change impacts. The conceptual framework and research design presented in this paper aim at helping to improve the empirical evidence of how people at the frontlines of climate change experience losses and damages. It brings a much-needed people-centred perspective into the Loss and Damage debate.

The COVID-19 pandemic made fieldwork-based empirical research in climate-vulnerable countries very difficult in the past few years (Hermans et al., 2021). With a new generation of young researchers, including many who are based in the Global South, focusing their attention on the emerging theme of Loss and Damage, it is my hope that some of the insights, shared in this paper, will help towards improving the quality of empirical, policy-relevant research in different parts of the world.

References

Adger, W. N. (1999). Social vulnerability to climate change and extremes in coastal Vietnam. *World Development*, 27(2), 249–269.

Afifi, T., Milan, A., Etzold, B., Schraven, B., Rademacher-Schulz, C., Sakdapolrak, P., … & Warner, K. (2016). Human mobility in response to rainfall variability: Opportunities for migration as a successful adaptation strategy in eight case studies. *Migration and Development*, 5(2), 254–274.

Ahmed, I., Ayeb-Karlsson, S., van der Geest, K., Huq, S., & Jordan, J. C. (2019). Climate change, environmental stress and loss of livelihoods can push people towards illegal activities: A case study from coastal Bangladesh. *Climate and Development*, 11(10), 907–917.

Ayeb-Karlsson, S. K., et al. (2016). A people-centred perspective on climate change, environmental stress, and livelihood resilience in Bangladesh. *Sustainability Science*, 11(4), 679–694.

Barnett, J., & S. O'Neill (2010). Maladaptation. *Global Environmental Change*, 20(2), 211–213.

Bekefi, T., Epstein, M. J., & Yuthas, K. (2008). *Managing opportunities and risks*. The Society of Management Accountants of Canada (CMA Canada), the American Institute of Certified Public Accountants, Inc. (AICPA) and The Chartered Institute of Management Accountants (CIMA).

Berkhout, F., & Dow, K. (2022). Limits to adaptation: Building an integrated research agenda. *Wiley Interdisciplinary Reviews: Climate Change*, e817.

Birkmann, J. (2011). First- and second-order adaptation to natural hazards and extreme events in the context of climate change. *Natural Hazards*, 58(2), 811–840.

Blaikie, P., et al. (1994). *At risk: Natural hazards, people's vulnerability, and disasters*. New York: Routledge.

Bouwer, L. M. (2019). Observed and projected impacts from extreme weather events: Implications for loss and damage. In *Loss and damage from climate change: Concepts, methods and policy options*, 63–82.

Boyd, E., Chaffin, B. C., Dorkenoo, K., Jackson, G., Harrington, L., N'guetta, A., ... & Stuart-Smith, R. (2021). Loss and damage from climate change: A new climate justice agenda. *One Earth*, 4(10), 1365–1370.

Burke Johnson, R., Onwuegbuzie, A. J., & Turner, L. A. (2007). Toward a definition of mixed methods research. *Journal of Mixed Methods Research*, 1(2), 112–133.

Carter, T. R., et al. (1994). *IPCC technical guidelines for assessing climate change impacts and adaptations*. London: Department of Geography, University College London.

Chambers, R. (1989). Editorial introduction: Vulnerability, coping and policy. *IDS Bulletin*, 20(2), 1–7.

Chandra, A., McNamara, K. E., Clissold, R., Tabe, T., & Westoby, R. (2023). Climate-induced non-economic loss and damage: Understanding policy responses, challenges, and future directions in pacific small island developing states. *Climate*, 11(3), 74.

Davies, S. (1996). *Adaptable livelihoods: Coping with food insecurity in the Malian Sahel*. London: MacMillan Press.

Dietz, T. (2013). 11 Participatory assessment of development in Africa. In *Local governance and poverty in developing nations* (pp. 216–240). Routledge.

de Sherbinin, A. (2014). Climate change hotspots mapping: What have we learned? *Climatic Change*, 123(1), 23–37.

Doelle, M., & Seck, S. (2019). Loss & damage from climate change: From concept to remedy? *Climate Policy*, 1630353. https://doi.org/10.1080/14693062.2019.

Dow, K., et al. (2013). Limits to adaptation. *Nature Climate Change*, 3(4), 305–307.

Ellis, F. (1998). Household strategies and rural livelihood diversification. *The Journal of Development Studies*, 35(1), 1–38.

Fankhauser, S., Smith, J. B., & Tol, R. S. (1999). Weathering climate change: Some simple rules to guide adaptation decisions. *Ecological Economics*, 30(1), 67–78.

Fankhauser, S., Dietz, S., & Gradwell, P. (2014). Non-economic losses in the context of the UNFCCC work programme on loss and damage (Policy Paper). London: Centre for Climate Change Economics and Policy, Grantham Research Institute on Climate Change and the Environment.

Füssel, H. M. (2007). Adaptation planning for climate change: Concepts, assessment approaches, and key lessons. *Sustainability Science*, 2(2), 265–275.

Füssel, H. M., & Klein, R. J. (2006). Climate change vulnerability assessments: An evolution of conceptual thinking. *Climatic Change*, 75(3), 301–329.

Hermans, K., Berger, E., Biber-Freudenberger, L., Bossenbroek, L., Ebeler, L., Groth, J., ... & Wiederkehr, C. (2021). Crisis-induced disruptions in

place-based social-ecological research-an opportunity for redirection. *GAIA-Ecological Perspectives for Science and Society*, 30(2), 72–76.
Intergovernmental Panel on Climate Change. (2007). *Contribution of working group II to the fourth assessment report of the Intergovernmental Panel on Climate Change*. Cambridge: Cambridge University Press.
Intergovernmental Panel on Climate Change. (2012). Glossary of terms. In Field, C. B., et al., eds. *Managing the risks of extreme events and disasters to advance climate change adaptation* (A Special Report of Working Groups I and II of the Intergovernmental Panel on Climate Change (IPCC)), 555–564. Cambridge and New York: Cambridge University Press.
Intergovernmental Panel on Climate Change. (2014). *Climate change 2014: Impacts, adaptation, and vulnerability. Contribution of working group 2 to the IPCC fifth assessment report*. http://www.ipcc-wg2.gov.
Intergovernmental Panel on Climate Change. (2022): Annex II: Glossary. In *Climate change 2022: Impacts, adaptation and vulnerability. Contribution of working group II to the sixth assessment report of the Intergovernmental Panel on Climate Change*, 2897–2930. Cambridge and New York: Cambridge University Press.
Janzen, S., Emerton, L., van der Geest, K., Narvaez, L., & Sebesvari, Z. (2021). Assessing losses and damages to ecosystem services: Current state and opportunities for the Warsaw International Mechanism under the UNFCCC. *Climate Policy, 21*(7), 912–926.
McLeman, R. (2010). Impacts of population change on vulnerability and the capacity to adapt to climate change and variability: A typology based on lessons from "a hard country". *Population and Environment*, 31(5), 286–316.
McNamara, K. E., & Jackson, G. (2019). Loss and damage: A review of the literature and directions for future research. *Wiley Interdisciplinary Reviews: Climate Change*, 10(2), e564.
Mechler, R., Singh, C., Ebi, K., Djalante, R., Thomas, A., James, R., ... & Revi, A. (2020). Loss and damage and limits to adaptation: Recent IPCC insights and implications for climate science and policy. *Sustainability Science*, 15, 1245-1251.
Morgan, D. L. (2007). Paradigms lost and pragmatism regained methodological implications of combining qualitative and quantitative methods. *Journal of Mixed Methods Research*, 1(1), 48–76.
Moser, S. C., & J. A. Ekstrom (2010). A framework to diagnose barriers to climate change adaptation. *PNAS*, 107(51), 22026–22031.
Opondo, D. O. (2013). Erosive coping after the 2011 floods in Kenya. *International Journal of Global Warming*, 5(4), 452–466.
Pouw, N., et al. (2016). Participatory assessment of development; lessons learned from an experimental approach in Ghana and Burkina Faso. *American Journal*

of Evaluation, 1–13. Advance online publication. https://doi.org/10.1177/1098214016641210.

Puig, D. (2022). Re-conceptualising climate change-driven 'loss and damage'. *International Journal of Global Warming*, 27(2), 202–212.

Rademacher-Schulz, C., Afifi, T., Warner, K., Rosenfeld, T., Milan, A., Etzold, B., & Sakdapolrak, P. (2012). *Rainfall variability, food security and human mobility: An approach for generating empirical evidence*. Intersections No. 10. Bonn: UNU-EHS.

Roberts, E., et al. (2014). Loss and damage: When adaptation is not enough. *Environmental Development*, 11, 219–227.

Scoones, I. (1998). Sustainable rural livelihoods: A framework for analysis (IDS Working Paper 72). Brighton: Institute of Development Studies.

Smit, B., et al. (2001) Adaptation to climate change in the context of sustainable development and equity. In McCarthy, J. J., et al., eds., *Climate change 2001: Impacts, adaptation and vulnerability*, IPCC Working Group II, 877–912. Cambridge: Cambridge University Press.

Tschakert, P., Ellis, N. R., Anderson, C., Kelly, A., & Obeng, J. (2019). One thousand ways to experience loss: A systematic analysis of climate-related intangible harm from around the world. *Global Environmental Change*, 55, 58–72.

UNFCCC. (2022). https://unfccc.int/news/cop27-reaches-breakthrough-agreement-on-new-loss-and-damage-fund-for-vulnerable-countries. Accessed on 20th November 2022.

van der Geest, K. (2004). *"We're Managing!" Climate change and livelihood vulnerability in Northwest Ghana*. Leiden: African Studies Centre.

van der Geest, K. (2011). *The Dagara farmer at home and away: Migration, environment and development in Ghana*. Leiden: African Studies Centre.

van der Geest, K., & Dietz, T. (2004). A literature survey about risk and vulnerability in drylands, with a focus on the Sahel. In Dietz, T., R. Rueben, R., & J. Verhagen, eds., *The impact of climate change on drylands*, 117–146. Dordrecht: Springer Netherlands.

van der Geest, K., & Schindler, M. (2016). *Case study report: Loss and damage from a catastrophic landslide in Sindhupalchok District, Nepal* (Report No. 17). Bonn: United Nations University Institute for Environment and Human Security (UNU-EHS).

van der Geest, K., & Schindler, M. (2017). *Handbook for assessing loss and damage in vulnerable communities* (Report No. 21). Bonn: United Nations University Institute for Environment and Human Security (UNU-EHS).

van Der Geest, K., & Van Den Berg, R. (2021). Slow-onset events: A review of the evidence from the IPCC Special Reports on Land, Oceans and Cryosphere. *Current Opinion in Environmental Sustainability*, 50, 109–120.

van der Geest, K., & Warner, K. (2015). Vulnerability, coping and loss and damage from climate events. In Collins, A. E., et al., eds., *Hazards, risks and disasters in society*, 121–144. Elsevier.

van der Geest, K., & Warner, K. (2020). Loss and damage in the IPCC Fifth Assessment Report (Working Group II): A text-mining analysis. *Climate Policy*, 20(6), 729–742.

Verheyen, R. (2012). Tackling loss and damage: A new role for the climate regime? Bonn: Germanwatch. http://loss-and-damage.net/4805.

Warner, K., & K. van der Geest (2013). Loss and damage from climate change: Local-level evidence from nine vulnerable countries. *International Journal of Global Warming*, 5(4), 367–386.

Zommers, Z., et al. (2016). *Loss and damage: The role of ecosystem services.* Nairobi: UNEP.

CHAPTER 6

Towards Effective Loss and Damage Systems in Disaster Recovery Contexts in Southeast Asia

Michael Boyland, Danny Marks, Ham Kimkong, and Agus Nugroho

Abstract This chapter examines research undertaken on the impact of disasters and climate change on development processes in Southeast Asia, which is considered one of the most vulnerable regions to disasters. The research investigates the disaster recovery processes and loss and

M. Boyland (✉)
SOAS, University of London, London, UK
e-mail: 703999@soas.ac.uk

D. Marks
School of Law and Government, Dublin City University, Dublin, Ireland

H. Kimkong
Department of Natural Resource Management and Development, Royal University of Phnom Penh, Phnom Penh, Cambodia

A. Nugroho
Stockholm Environment Institute, Bangkok, Thailand

V. Wijenayake et al. (eds.), *Linking Climate Change Adaptation, Disaster Risk Reduction, and Loss & Damage*,
https://doi.org/10.1007/978-981-99-8055-0_6

damage systems in four long-term disaster contexts across Southeast Asia, including floods in the Vietnamese Mekong Delta, livelihood and infrastructure flood responses in Prey Veng, Cambodia, traditional fisheries management in tsunami recovery in Aceh, Indonesia and SMEs recovery from floods in Central Thailand. The study develops a typology of loss and damage systems and identifies effective intervention strategies, both formal and informal, to achieve more harmonized disaster recovery and climate loss and damage systems. A harmonized approach in disaster and climate governance is essential to address the root causes of risk, instead of just the impacts and consequences of ongoing disaster events and climatic processes.

Keywords Disasters · Loss and damage · Resilience · Disaster recovery · Southeast Asia

Highlights

- The distinctions between loss and damage and disaster recovery approaches remain disputed.
- Qualitative analysis of four major disaster recovery contexts in Southeast Asia of the past two decades to examine loss and damage systems of different types.
- A more holistic and connected approach to addressing disaster impacts and climate losses and damages could enable the root causes of risk and vulnerability to be addressed.

1 Introduction

Southeast Asia is a region at risk to the impacts of disasters and climate change. Disasters have had significant social, economic and environmental consequences for decades (Thomalla et al. 2017a), and large parts of the region are exposed and sensitive to changes in climate-related hazard frequency, intensity and distribution driven by climate change (ASEAN 2021). Besides the exposure to a range of hazards, disaster and climate risk levels are also influenced by vulnerability and its root causes; high rates of poverty and inequality, rapid and unplanned urbanization, natural resource degradation, weak governance institutions, and non-risk-informed policies—as described in the Sendai Framework for Disaster Risk Reduction (UNGA 2015)—plague the region (IPCC 2022; UNDRR 2022).

The effective management and reduction of disaster and climate risks is a priority for the Association of Southeast Asian Nations (ASEAN); the regional intergovernmental body consisting of ten ASEAN Member States (AMS). However, efforts have been hampered by a short-term, 'ex-post' perspective that prioritizes managing and responding to events over addressing root causes of risk and vulnerability. The additional risk layer of climate change further complicates the challenge, and the governance of disasters and climate change remains a fragmented process at most levels, particularly for floods (Chau et al. 2014; Lebel and Lebel 2017; Boyland 2019). Insufficient and ineffective disaster risk reduction (DRR), coupled with a siloed approach to adapt to climate change and address loss and damage (L&D), can lead to more frequent and worsening disaster and extreme events, with complex and prolonged recovery phases.

1.1 Framing Loss and Damage in Disaster and Climate Contexts

The United Nations Framework Convention on Climate Change (UNFCCC) defines loss and damage (L&D) as 'the actual and/or potential manifestation of impacts associated with climate change in developing countries that negatively affect human and natural systems' (UNFCCC 2012). In the context of international climate governance, L&D is an important aspect of climate action which at its core concerns the compensation for impacts in climate vulnerable lower-income countries by industrialized nations responsible for historic emissions (Calliari 2016). This origin as a highly political and contentious issue between developed and developing countries in climate negotiations has evolved over the last decade to become an important pillar of the climate policy discourse post-Paris Agreement adoption (i.e. at COP21 in 2015), particularly as the limits of adaptation are tested (Vulturius and Davis 2016). Somewhat ambiguously, however, the Paris Agreement calls for 'action and support' on loss on damage, rather than compensation. Since then, negotiators and observers have been frustrated with the slow progress and lack of political and financial commitment to address L&D from wealthy nations (Chandrasekar 2017; Serdeczny 2017). As a result, L&D has become a major talking point of climate negotiations and progress on Paris Agreement implementation.

The UNFCCC framing leaves aspects open to interpretation and the concept has evolved to resonate in different ways with different climate change, disasters, and humanitarian communities (Mechler et al. 2014; Boyd et al. 2017; James et al. 2014). As a result, a range of broader framings of L&D have been considered by research and practice communities, including those working on adaptation and DRR (James et al. 2014; Boyd et al. 2017). Recognizing this trend in the L&D discourse, Boyd et al. (2017) identify a range, or continuum, of framings (see Fig. 1). The *risk management* framing reflects on connected L&D, adaptation, and DRR approaches. Disaster management and recovery initiatives complement loss and damage systems which aim to address 'actual loss and damage' rather than prevent 'potential loss and damage' (Boyd et al. 2017).

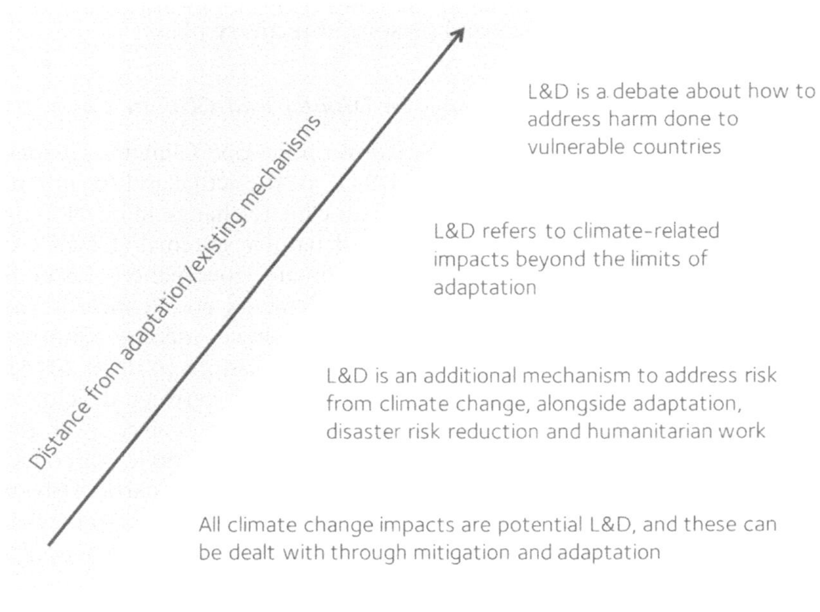

Fig. 1 Loss and damage (L&D) framings and their distance from adaptation

In this paper, we utilize the term 'loss and damage systems', which we define as the formal and informal institutions, processes, and systematic actions which aim to assist communities and societies to cope with, adapt to, and recover from adverse effects of disasters and climate change that may be either irreversible (loss) or replaceable (damage) (see Thomalla et al. 2017b). This builds on framings and definitions of both L&D and disaster recovery (UNGA 2016; UNFCCC 2012). As efforts to reduce climate loss and damage move from the negotiation halls to the communities on the frontline of climate change, a better understanding of the performance of loss and damage systems in the context of broader risk management approaches is needed to cope with and adapt to the long-term impacts of climate change (Surminski and Lopez 2014; Thomalla et al. 2017b).

This article aims to understand the loss and damage systems present in different major, long-term disaster contexts across Southeast Asia. Our analysis of the following four empirical case studies in Cambodia, Thailand, Vietnam and Indonesia will examine how separate DRR and L&D approaches can become more integrated:

1. Strategies for living with seasonal floods and livelihood recoveries of farmers in An Giang province, in the Vietnamese Mekong Delta (VMD), following the 2000 flood;
2. Local recovery approaches and residential cluster schemes to build resilience in Prey Veng province, Cambodia, following the 2000–2001 flood;
3. The role of a fisheries customary institution (*Panglima Laot*) in the recovery of coastal communities in Aceh, Indonesia, following the 2004 Indian Ocean tsunami; and
4. The economic recovery of small and medium enterprises (SMEs) in Bang Bua Thong (BBT) market, Nonthaburi province, Thailand, following the 2011 flood event.

In the context of these case studies, we seek to answer three research questions:

1. What role do loss and damage systems play in disaster recovery contexts?

2. What lessons for improving loss and damage systems can be learned from major disaster recoveries?
3. How can loss and damage systems in disaster recovery contexts be more effective and harmonized with DRR approaches and practices?

2 Methodology

In this study, conducted between 2014 and 2018, we take a qualitative methodological approach. Firstly, we undertook secondary data analysis of academic and grey literature, and government/policy documents, related to disasters, disaster recovery, climate change adaptation, loss and damage and post-disaster development. Primarily English-language research literature was gathered from Scopus and Google Scholar search engines: we developed keyword search strings and screened the titles and abstracts of all results, narrowing down relevant papers which were reviewed in full (see Thomalla et al. 2017b for further detail). This informed our understanding of key topics and shaped our framework for linking post-disaster and climate loss and damage systems. Policies and other government or national civil society documents written in Thai, Bahasa Indonesian, Khmer, and Vietnamese were reviewed by the respective research teams to inform our understanding of each disaster context in Thailand, Indonesia, Cambodia, and Vietnam, respectively. The research team convened periodically to share results and discuss the research design, which led to the finalization of our primary data collection approach, including ethics procedures and interview protocols adapted for each country context.

Secondly, we conducted between 20 and 25 interviews in each case study country with relevant actors and stakeholders such as officials and planners in national and local governing authorities; representatives of donors, aid agencies, research institutions, international and national NGOs, and local businesses; and members of disaster-affected communities. In each country, the number of interviews and split between representatives of bodies and groups differed depending on factors including the disaster recovery context, the amount of time passed since the disaster, and the availability of individuals. However, in each case, the interviews were semi-structured, aligned with our research framework, and conducted in local languages before being transcribed and translated into English—an approach designed and implemented by the research team (see Thomalla et al. 2017b for further detail). Through a series of workshops among the research team, the qualitative analysis was consistent across all primary data.

2.1 Case Studies

Each case study is a focused view on specific aspects of a larger-scale recovery process, several of which spanned multiple countries, including the 2004 Indian Ocean tsunami and the 2000–2001 Mekong Delta floods (see Table 1). While the cases vary in terms of scale of impact, recovery period, and development context, we analyse and compare prevalent loss and damage systems, paying attention to target beneficiaries, and how they were planned and implemented.

Adapted from Thomalla et al. (2017b) and with disaster data from CRED and Guha-Sapir (2017).

The Vietnam case study examines An Giang farmers' coping and adaptive capacity following extreme seasonal flooding in 2000. An Giang province is flood-prone due to a various natural and man-made factors (Le et al. 2007). The notion of 'living with floods' became a national government strategy following the floods, prioritizing building 'flood-proof' houses, adjusting agricultural schedules, and enhancing flood release capacity in the Delta area (Tinh and Hang 2003). The VMD is essential to the food security and development of the region but is an area with high-flood risk exposure and vulnerability. Without adaptation and risk management, risk levels will increase as the sea level rises, land-use changes, and urbanization rates accelerate (Garschagen et al. 2011).

The Cambodia case study concerns resilience programmes and residential cluster schemes as an adaptation strategy in Prey Veng, following extreme floods in 2000 and 2001. Flood impacts were concentrated in areas along the Mekong River and around Tonle Sap Lake, destroying houses, infrastructure, farmland, and livelihoods, resulting in food insecurity for at least one million people. The humanitarian response by government and NGOs focused on short-term emergency aid. To tackle the recovery and long-term resilience building, the government established the National Committee for Disaster Management (NCDM) and equivalent bodies at provincial and district levels.

The Indonesia case study analyses the roles played by the *Panglima Laot*, a fisheries customary institution, in the recovery of coastal communities in Aceh following the 2004 Indian Ocean tsunami. The human and environmental impacts of the major disaster are well documented, as

Table 1 Overview of the four case studies, including key loss and damage data (at the national level)

Case study	*Year*	*Disaster type*	*Deaths (national)*	*Total people affected*	*Economic losses and damage (US$)*
'Living with flood' agricultural strategies in An Giang, Vietnam	2000	Flood	496	5 million	265 million
Local recovery and residential cluster schemes in Prey Veng, Cambodia	2000–2001	Flood	403	5 million	175 million
Role of *Panglima Laot* in the recovery in Aceh, Indonesia	2004	Tsunami	165,800	700,000	4.5 billion
Recovery of small and medium enterprises (SMEs) in Nonthaburi, Thailand	2011	Flood	877	10 million	40 billion

are the political events in Aceh in the months after the disaster, which saw the end of decades of civil conflict when the Indonesian government and the Free Aceh Movement signed a peace deal (Telford and Cosgrave 2006; Fan 2013). Due to the scale of the disaster, the response and recovery phases lasted years and had fundamental impacts on development outcomes in Aceh.

The Thailand case study considers how small businesses in Bang Bua Tong (BBT), a market area in Nonthaburi, were impacted by and able to recover from the 2011 flood, which has widespread impacts (CRED and Guha-Sapir 2017; World Bank 2012). SMEs are essential to the Thai economy; approximately 2.7 million businesses contribute 37 per cent gross domestic product (GDP) and account for 80 per cent of the workforce (ADB 2014). In BBT, economic impacts were severe and have persisted for years, compounded by a weakening national economy (Marks and Thomalla 2017).

3 Results: Loss and Damage Systems

We identify a total of 14 different loss and damage system initiatives across the cases, categorized in five domains—livelihoods, infrastructure, financial, institutional, and ecological. The initiatives are identified through data collection, while the domains align with operational disaster recovery frameworks (FEMA 2017; GFDRR 2015). Table 2 summarizes the loss and damage system initiatives we found in each case.

3.1 Livelihoods

Livelihood and economic sector stimulus policies were drafted and implemented by national and local governments in all cases. In Thailand, macro-economic policies included lowering interest rates, giving tax breaks on foreign investments, and a new car tax rebate for first-time car buyers to revive the heavily impacted car manufacturing sector. Despite such initiatives, SMEs felt they were left behind and de-prioritized in receiving livelihood assistance compared with other sectors, and transnational companies were able to lobby for certain measures to aid their recovery. In Indonesia, lost and damaged fishing boats and equipment were replaced relatively quickly, although in certain cases the fisherfolk were not consulted and unsuitable boats were provided.

Table 2 Loss and damage system initiatives present in each case study, categorized by five main disaster recovery sectors

Recovery sector	Loss and damage system initiative	'Living with floods', An Giang, Vietnam	Residential cluster schemes, Prey Veng, Cambodia	The role of Panglima Laot, Aceh, Indonesia	SMEs economic recovery, Nonthaburi, Thailand
Livelihoods	Rehabilitation	X		X	
	Stimulus				X
	Transformation	X	X	X	
Infrastructure	Reconstruction		X	X	X
	Protection	X		X	X
	Shelter and housing	X	X	X	X
Financial	Compensation		X	X	X
	Insurance				X
	Community funds		X	X	
	Loans and micro-credit	X	X		
Institutional	Professional associations				
	Local institutions				
Ecological	DRR ecosystem services				
	Livelihood ecosystem services				

The 'living with floods' approach introduced an additional annual rice harvest and alternative off-farm livelihood initiatives for additional income during floods in Vietnam. Similarly, in Cambodia, training for non-agricultural livelihoods aimed to support alternative income-generating activities, but economic benefits were uneven between participating households and some become economic migrants to urban areas. In BBT, SMEs tended to revert to pre-flood activities, i.e. selling the same goods and offering the same services. Changes in the socio-political context in Aceh constituted a transformation for development outcomes, including property ownership and tenure rights for women and other groups.

3.2 Infrastructure

The infrastructure recovery category includes initiatives for reconstruction, protection, and housing. In Indonesia, a dedicated reconstruction fund was established, a coastal exclusion zone created (although this decision was later partly reversed), some reconstruction and road building and upgrading conducted further inland, and a Tsunami Early Warning System (including evacuation routes and evacuation centres) established. For resilience-building in Thailand, a newly drafted Flood Management Plan prioritized hard infrastructure measures including flood defences along waterways, raised roads and walls around industrial estates, over holistic changes to land-use planning. In Cambodia, Indonesia and Vietnam, upgrading roads and bridges for community protection and improved access to jobs, education and services was a central policy. For SMEs in Thailand, shop floor levels were raised, and stock, equipment, and offices were moved upstairs during monsoon seasons. In Cambodia, the residential cluster programme introduced a new house-building approach which prioritized flood adaptation and some residents re-built at high levels than pre-flood. Thousands of brand-new homes were built by both government and non-government entities in Aceh.

3.3 Financial

Financial loss and damage initiatives were evident in all cases and ranged from insurance to micro-credit. In BBT, uneven levels of compensation led to protests and calls for transparency and equity in the distribution of support, particularly as so few SMEs had insurance. Cash-for-work drives occurred in Cambodia and Indonesia. In Vietnam, compensation for affected households was available for a death or injury in the family, for property damages and for agricultural losses. Community-level funds and micro-credit schemes for household and business repairs were important loss and damage systems in Vietnam, Cambodia and Indonesia, particularly as access to bank loans for reconstruction or land purchasing were often hard to access. In Indonesia, the *Panglima Laot* fisheries customary institution worked with INGOs to establish a revolving fund for fisheries communities to access for the purchase of livelihood equipment and supplies.

3.4 Institutional

Community-based institutions played important roles in supporting recoveries by providing various loss and damage system initiatives. In Cambodia, a disaster association provided material assistance such as food and building materials. In Indonesia, the *Panglima Laot* (operating at community, district and provincial levels) gave a voice to affected communities, for example, in the BRR-coordinated recovery decision-making processes and advocated for the needs and priorities of the fisherfolk. In Thailand, the market association prepared community disaster preparedness plans. In Vietnam, local institutions were the first line of support as floods began, under the 'four on-the spot' disaster management policy *(Phuong cham 4 tai cho)* which mobilizes local response support. Local knowledge and experience with weather conditions and disaster indicators were important to preparedness and initial responses in all cases. In Indonesia, the principle of 'gotong royong' (mutual cooperation) meant communities were proactive and grassroots efforts sprang up, although INGOs who came to Aceh with their own ideas, intentions, and procedures were not always able to work efficiently with these groups.

3.5 Ecological

The provision of ecosystem services for adaptation, DRR, and livelihoods were found in all cases. In Thailand, natural floodwater retention areas, i.e. wetlands and retention ponds, were not in place in the urban areas affected by the floods in 2011. Since then, they are being implemented, but with the aim of reducing water shortages during droughts. The principles of 'living with floods' in Vietnam are around accommodating rather than controlling flood waters, for example, through the use of natural semi-dykes which allow the replenishment of soil nutrients. In Indonesia, there was a trade-off between the protection of marine ecosystems and the recovery of the fisheries sector, complicated by unsustainable fishing practices and depleted fish stocks even prior to the disaster, and so the socio-ecological issue was at times overlooked as a component of the disaster recovery.

4 Discussion

4.1 Improving Loss and Damage Systems in Disaster Recovery Contexts

The case study results have identified loss and damage systems initiatives present in major disaster recovery processes in different contexts in Southeast Asia. Drawing on these results and supplementary literature, we discuss limitations and barriers in the governance of loss and damage systems and suggest opportunities to overcome them.

First, there was limited downward accountability of loss and damage systems in all cases. One reason for this is that governance structures in these countries remain centralized. National governments have retained the majority share of decision-making power and resources in order to protect national interests. Despite this, national government agencies can become overwhelmed and respond ineffectively and inefficiently following a major disaster, such as in the wake of the 2004 Indian Ocean Tsunami (Boyland et al. 2017) and the 2011 flood event in Thailand (Marks and Thomalla 2017).

Second, fragmented governance arrangements of loss and damage systems in recovery contexts limit collaboration and even cause unintended negative consequences of attempted recovery interventions. One reason for this is because individual agencies seek to protect their mandates and budgets (Marks and Lebel 2016). Participation is a key facet of ensuring accountability and maintaining transparency, but government agencies tend to exclude key groups as both recovery actors and beneficiaries in some contexts. For example, SMEs were largely sidelined despite the major role they play in Thailand's local and national economies. The delegation of recovery implementation responsibilities to sub-national levels is recommended for effectiveness and efficiency (GFDRR 2015).

Thirdly, the agency capacity and resources for working on disaster management remains insufficient, with a general focus on short-term disaster response action rather than long-term recovery and resilience-building approaches. As a result, limited financial and human resources have been devoted to multi-hazard, comprehensive DRR that connects with broader development efforts.

Given such limitations, how can these systems be improved? We suggest six recommendations. First, legal foundations of DRR policies and practices need to be more institutionalized since they currently create uncertainty regarding agencies' DRR responsibilities in these countries, which particularly come to light during complex, high-pressure recovery phases (Chau et al. 2014; Marks and Lebel 2016). Second, a more devolved, participatory, and cross-level decision-making system that engages all levels of government as well as external agencies and NGOs is needed. Disaster-affected people whose disaster risk and development futures are being shaped by loss and damage system initiatives should be central to post-disaster decision-making processes. Third, multi-agency and multi-scale inclusion needs to be ensured in ways that avoid duplication of effort. Overlaps in mandates, particularly in the water resource management (WRM) sector, are common in the region (Lebel et al. 2005) and need to be minimized for the sake of effective disaster recovery governance. Fourth, while agencies and practitioners are encouraged to incorporate 'building back better' principles in recovery visions and frameworks, more needs to be done to advance understanding of the term to go beyond 'principles' and in practice to represent more than reconstruction of infrastructure and housing. A heavily structural-focused recovery strategy may simply redistribute rather than reduce risk (Lebel and Sinh 2009). Fifth, clear guidelines, milestones, and agency responsibilities for transitioning from recovery to development are required in major disaster recovery contexts, particularly in terms of feeding into existing land-use planning and WRM policy. Development policy must also be flexible and adaptive to anticipate major disaster recoveries, which may shift development priorities, possibly even for the better. Sixth and finally, considering these limitations in recovery governance, disaster-affected communities should not remain passive but rather they should self-organize to provide social safety nets to those worst affected, become active voices in planning processes, create their own early warning systems and advocate for targeted loss and damage assistance during disaster recoveries. Table 3

Table 3 Summary of limitations and their consequences in the governance of loss and damage systems in disaster recovery contexts, as well as recommendations to address them

Limitation	*Consequences*	*Recommendations*
Limited accountability and transparency	• Inequitable interventions and support • Nepotism and corruption • No long-term responsibilities • Limited monitoring and evaluation	• Target assistance and safety nets to most vulnerable • Independent transparency and accountability mechanisms • Balance trade-off between accountability and bureaucracy • Institutionalize monitoring & evaluation, long-term reviews
Fragmented governance arrangements	• Poor coordination and collaboration • Weak participation in decision-making • Ineffective planning • Services not provided in timely fashion	• Utilize inter-agency task forces and coordination bodies with clear mandates • Ensure participation in decision-making processes • Leverage ongoing development initiatives for recovery (e.g., social services) • Support self-organizing, grassroots initiatives

summarizes the discussed limitations and opportunities for loss and damage system governance in recovery contexts.

4.2 *Towards Integrated Loss and Damage Systems and Disaster Recovery Approaches*

The impacts of disasters, as demonstrated here, can be compounded by insufficient loss and damage systems and ineffective disaster recovery approaches. When disaster management and governance systems are not equipped to cope with and adapt to disaster events of a magnitude that exceeds expected or 'normal' levels, people suffer deeper consequences.

The post-disaster phases of response and recovery are critical for determining how severely and for how long a disaster will impact people. Our results highlight that loss and damage systems play an important role in disaster recoveries, but they tend to be thought of as distinct approaches from one another. Here, we discuss the differences and commonalities between disaster recovery interventions and loss and damage systems, and consider options for greater harmonization of the two approaches.

The first point of divergence for loss and damage and disaster recovery approaches comes at the international policy level. As noted, the concept of loss and damage is rooted in international, UNFCCC-framed climate change negotiations and mechanisms. Disaster recovery, meanwhile, is arguably a less political-driven term that is more rooted in experienced impacts of disasters and efforts to recover from lost and damaged assets to a state of normality again. At all levels, climate change and disaster management governance are typically divorced from one another. At the global level, the UNFCCC orchestrates legally binding climate change policy for all parties (i.e. the Paris Agreement), while UNDRR is unable to secure the same level of commitment from nations to tackle disaster risks; the 2015-adopted Sendai Framework for DRR is a non-binding policy framework document. Further, the two policy processes do not sufficiently recognize one another as complementary and working towards mutual goals in the broader context of the Sustainable Development Goals (SDGs). In ASEAN, AMS governance arrangements are similarly fragmented and siloed; climate change is typically framed and governed as an environmental issue, while flood risk management is a water sector concern (e.g., Chau et al. 2014; Marks and Lebel 2016; Lebel and Lebel 2017). Recovery actors must improve integration and collaboration beyond the limits of institutional arrangements and mandates. A priori recovery institutionalization is a key enabler of effective processes and interventions (GFDRR 2015).

In recent years, 'build back better' has emerged as a widely adopted goal of disaster recoveries—originally articulated by former US President Bill Clinton in the aftermath of the 2004 Indian Ocean Tsunami (in his capacity as UN Special Envoy for Tsunami Recovery) (Fan 2013). Although building back better—using a disaster as an opportunity to improve societies—is clearly a desirable goal, some argue 'build back better' is a narrative owned by international interests that have politicized disaster recoveries at the expense of the empowerment of those affected, who can have very different priorities in recovery (Fan 2013; Thomalla et al. 2017b; Boyland et al. 2017). This also links to the broader counter-narrative of disaster capitalism by the state and private actors (Loewenstein 2015; Klein 2007). Disaster recovery and loss and damage actors alike can work together to ensure that post-disaster initiatives follow inclusive, people-centred approaches and decision-making processes where the power is not concentrated in the hands of a few officials.

While we are not arguing for a reinvention of the wheel, given the layer of risk that climate change adds to disaster risks, the DRR community, including recovery actors, must take account of these changing risks in their approaches. Disaster recovery approaches are typically short- and medium-term in nature, but would be enhanced by taking a longer-term view of disasters and climate change, and seeking to better link with resilience and development actions (Mosel and Levine 2014; Thomalla et al. 2017b). GFDRR (2015) suggests that linked resilient recovery and development integrates gender equity, vulnerability reduction, natural resource conservation, environmental protection and climate change adaptation. Boyd et al. (2017) argue the framing of loss and damage can facilitate better integration of adaptation and DRR approaches. Common ground for such integration may lie in realigning both approaches to holistically tackling the root causes of risk that are generated within unsustainable development processes (UNDRR 2022; Thomalla et al. 2018).

5 Conclusions

This article has analysed loss and damage systems in disaster recovery processes following four major events in Southeast Asia. Through empirical case study analysis, we have also highlighted where disaster recovery and loss and damage approaches to disaster and climate risk reduction converge and diverge. A diversity of systems and interventions were found, in terms of livelihoods, infrastructure, financial, institutional, and ecological recovery sectors. This reflects both the severity of the disaster impacts and the diversity of actors working in different loss and damage systems as part of major disaster recoveries in each of the four case studies.

While important support and services were provided to affected communities in each of the cases, loss and damage systems can more effectively build resilience by considering the underlying drivers of vulnerability, how to equitably reach intended beneficiaries and societal groups (e.g., women, children, and persons with disabilities), and overarching governance constraints such as the effectiveness, coordination, and accountability of national and local authorities together with international and national NGOs.

Loss and damage systems and disaster recovery interventions have several commonalities and differences, but the results of the case studies suggest that the approaches could be pursued in greater unison. Both approaches can learn lessons from one another in order to ensure that climate and disaster risks are addressed in harmony, by taking a multi-risk and longer-term approach when recovering from disasters. The Southeast Asia region is a global 'hot spot' for climate and disaster risks, and actors at all levels need to address the conflicts created by fragmented institutional arrangements and power imbalances to ensure the long-term resilience and sustainability of the region's people and environments.

Acknowledgements The research extends its gratitude to the Asia Pacific Network for Global Environmental Change Research (APN) for supporting the project. We also extend our thanks to numerous individuals who assisted with field data collection, interviews, or provided valuable information for the project. Special appreciation goes to Suttirak San-ngah, Sok Sophoarn, Dton Siriwan, Dr. Nguyen Duy Can, Nguyen Quynh Anh, and Do Thuy Ngan.

References

ADB. 2014. *Asia SME Monitor 2013*. Manila: Asian Development Bank (ADB).

ASEAN. 2021. *ASEAN Disaster Resilience Outlook: Preparing for a Future Beyond 2025*. Jakarta: ASEAN Secretariat. https://asean.org/wp-content/uploads/2021/10/ASEAN-Disaster-Resilience-Outlook-Preparing-for-the-Future-Beyond-2021-FINAL.pdf.

Boyd, Emily, Rachel A. James, Richard G. Jones, Hannah R. Young, and Friederike E. L. Otto. 2017. "A Typology of Loss and Damage Perspectives." *Nature Climate Change* 7 (10): 723–29. https://doi.org/10.1038/nclimate3389.

Boyland, M. 2019. *In Pursuit of Effective Flood Risk Management in the Mekong Region*. SEI Discussion Brief. Bangkok: Stockholm Environment Institute (SEI). https://www.sei.org/publications/in-pursuit-of-effective-flood-risk-management-in-the-mekong-region/.

Boyland, Michael, Agus Nugroho, and Frank Thomalla. 2017. "The Role of the Panglima Laot Customary Institution in the 2004 Indian Ocean Tsunami Recovery in Aceh." In *Disaster Risk Reduction in Indonesia*, edited by Riyanti

Djalante, Matthias Garschagen, Frank Thomalla, and Rajib Shaw, 357–76. Cham: Springer International Publishing. https://doi.org/10.1007/978-3-319-54466-3_14.

Calliari, E (2016). Loss and damage: A critical discourse analysis of Parties' positions in climate change negotiations. *Journal of Risk Research* 21 (6): 725–747. https://doi.org/10.1080/13669877.2016.1240706.

Chandrasekar, Aruna. 2017. "COP23: Loss and Damage Text Analysis." *COP23: Loss and Damage Text Analysis* (blog). November 15. http://climatetracker.org/cop-23-loss-damage-text-analysis/.

Chau, Vu Ngoc, John Holland, and Sue Cassells. 2014. "Institutional Structures Underpinning Flood Management in Vietnam." *International Journal of Disaster Risk Reduction* 10 (December): 341–48. https://doi.org/10.1016/j.ijdrr.2014.10.008.

CRED, and D. Guha-Sapir. 2017. *EM-DAT: The Emergency Events Database—Université Catholique de Louvain (UCL)*. Centre for Research on the Epidemiology of Disasters (CRED). http://www.emdat.be.

Fan, L. 2013. *Disaster as Opportunity? Building Back Better in Aceh, Myanmar and Haiti*. Overseas Development Institute.

FEMA. 2017. *National Disaster Recovery Framework*. Washington, D.C.: U.S. Department of Homeland Security—Federal Emergency Management Agency (FEMA). https://www.fema.gov/pdf/recoveryframework/ndrf.pdf.

Garschagen, Matthias, Fabrice G. Renaud, and Jörn Birkmann. 2011. "Dynamic Resilience of Peri-Urban Agriculturalists in the Mekong Delta Under Pressures of Socio-Economic Transformation and Climate Change." In *Environmental Change and Agricultural Sustainability in the Mekong Delta*, 141–63. Advances in Global Change Research. Dordrecht: Springer. https://doi.org/10.1007/978-94-007-0934-8_9.

GFDRR. 2015. *Disaster Recovery Framework—Guide*. Washington, D.C.: Global Facility for Disaster Reduction and Recovery (GFDRR).

IPCC. 2022. "Climate Change 2022: Impacts, Adaptation, and Vulnerability. Contribution of Working Group II to the Sixth Assessment Report of the Intergovernmental Panel on Climate Change." Cambridge University Press, IPCC.

James, Rachel, Friederike Otto, Hannah Parker, Emily Boyd, Rosalind Cornforth, Daniel Mitchell, and Myles Allen. 2014. "Characterizing Loss and Damage from Climate Change." *Nature Climate Change* 4 (11): 938–39. https://doi.org/10.1038/nclimate2411.

Klein, Naomi. 2007. *The Shock Doctrine: The Rise of Disaster Capitalism*. 1st ed. Toronto: Knopf Canada.

Le, Thi Viet Hoa, Huu Nhan Nguyen, Eric Wolanski, Thanh Cong Tran, and Shigeko Haruyama. 2007. "The Combined Impact on the Flooding in Vietnam's Mekong River Delta of Local Man-Made Structures, Sea Level Rise, and Dams Upstream in the River Catchment." *Estuarine, Coastal and Shelf Science* 71 (1–2): 110–16. https://doi.org/10.1016/j.ecss.2006.08.021.

Lebel, Louis, and Bach Tan Sinh. 2009. "Risk Reduction or Redistribution? Flood Management in the Mekong Region." *Asian Journal of Environment and Disaster Management (AJEDM)—Focusing on Pro-Active Risk Reduction in Asia* 1 (1): 25. https://doi.org/10.3850/S179392402009000040.

Lebel, Louis, and Phimphakan Lebel. 2017. "Policy Narratives Help Maintain Institutional Traps in the Governance of Floods in Thailand." *International Journal of Water Resources Development*, July, 1–16. https://doi.org/10.1080/07900627.2017.1336989.

Lebel, Louis, Po Garden, and Masao Imamura. 2005. "The Politics of Scale, Position, and Place in the Governance of Water Resources in the Mekong Region." *Ecology and Society* 10 (2): 18.

Loewenstein, A. 2015. *Disaster Capitalism: Making a Killing Out of Catastrophe.* London, and Brooklyn, NY: Verso.

Marks, Danny, and Frank Thomalla. 2017. "Responses to the 2011 Floods in Central Thailand: Perpetuating the Vulnerability of Small and Medium Enterprises?" *Natural Hazards* 87 (2): 1147–65. https://doi.org/10.1007/s11069-017-2813-7.

Marks, Danny, and Louis Lebel. 2016. "Disaster Governance and the Scalar Politics of Incomplete Decentralization: Fragmented and Contested Responses to the 2011 Floods in Central Thailand." *Habitat International*, 52 (March): 57–66. https://doi.org/10.1016/j.habitatint.2015.08.024.

Mechler, Reinhard, Laurens M. Bouwer, Joanne Linnerooth-Bayer, Stefan Hochrainer-Stigler, Jeroen C. J. H. Aerts, Swenja Surminski, and Keith Williges. 2014. "Managing Unnatural Disaster Risk from Climate Extremes." *Nature Climate Change* 4 (4): 235–37. https://doi.org/10.1038/nclimate2137.

Mosel, I., and S. Levine. 2014. *Remaking the Case for Linking Relief, Rehabilitation and Development. How LRRD Can Become a Practically Useful Concept for Assistance in Difficult Places*. London: Overseas Development Institute.

Serdeczny, Olivia. 2017. "Loss and Damage at COP23—Goals, Roadblocks and Detours." *Climate Analytics*. November 20. http://climateanalytics.org/blog/2017/loss-and-damage-at-cop23-disappointments-and-consolation-prizes.html.

Surminski, Swenja, and Ana Lopez. 2014. "Concept of Loss and Damage of Climate Change—A New Challenge for Climate Decision-Making? A Climate Science Perspective." *Climate and Development* 1–11. https://doi.org/10.1080/17565529.2014.934770.

Telford, J., and J. Cosgrave. 2006. *Tsunami Evaluation Coalition—Joint Evaluation of the International Response to the Indian Ocean Tsunami*. Synthesis Report. London: Tsunami Evaluation Coalition.

Thomalla, F., M. Boyland, K. Johnson, et al. 2018. "Transforming Development and Disaster Risk." *Sustainability*.

Thomalla, Frank, Michael Boyland, and Emma Calgaro. 2017. "Disasters and Development in Southeast Asia Toward Equitable Resilience and Sustainability." In *Routledge Handbook of Southeast Asian Development*, by Lisa Law and Fiona Miller, edited by Andrew McGregor, 1st ed., 342–61. New York: Routledge. https://doi.org/10.4324/9781315726106-30.

Thomalla, Frank, Louis Lebel, Michael Boyland, Danny Marks, Ham Kimkong, Sinh Bach Tan, and Agus Nugroho. 2017. "Long-Term Recovery Narratives Following Major Disasters in Southeast Asia." *Regional Environmental Change*, December, 1–12. https://doi.org/10.1007/s10113-017-1260-z.

Tinh, D. Q., and P. T. Hang. 2003. "Living with Floods in the Mekong River Delta of Vietnam." Paper for 3 World Water Forum—Poverty and Flood Theme. Vietnam.

UNDRR. 2022. *Global Assessment Report for Disaster Risk Reduction*. Geneva: United Nations Office for Disaster Risk Reduction (UNDRR).

UNFCCC. 2012. *A Literature Review on the Topics in the Context of Thematic Area 2 of the Work Programme on Loss and Damage: A Range of Approaches to Address Loss and Damage Associated with the Adverse Effects of Climate Change*. FCCC/SBI/2012/INF.14. Doha: UNFCCC.

UNGA. 2015. *Sendai Framework for Disaster Risk Reduction 2015–2030*. A/CONF.224/CRP.1. Sendai: UN.

UNGA. 2016. *Report of the Open-Ended Intergovernmental Expert Working Group on Indicators and Terminology Relating to Disaster Risk Reduction*. New York: United Nations General Assembly. https://www.preventionweb.net/publications/view/51748.

Vulturius, Gregor, and Marion Davis. 2016. *Defining Loss and Damage: The Science and Politics around One of the Most Contested Issues within the UNFCCC*. Discussion Brief. Stockholm: Stockholm Environment Institute.

World Bank. 2012. "Thai Flood 2011: Rapid Assessment for Resilient Recovery and Reconstruction Planning (Vol. 2 of 2): Final Report (English)." Thai Flood 2011 : Rapid Assessment for Resilient Recovery and Reconstruction Planning. Bangkok: World Bank.

CHAPTER 7

Identification of Non-economic Loss and Damage (NELD) Indicators and Practices in the Context of Climatic Events

Sivapuram Venkata Rama Krishna Prabhakar, Yohei Chiba, and Md. Atikul Islam

Abstract Loss and damages (L&D) caused by climate change have become important in international and national discussions. However, the least understood and addressed aspect of L&D is non-economic loss

S. V. R. K. Prabhakar (✉)
Institute for Global Environmental Strategies, Hayama, Japan
e-mail: prabhakar@iges.or.jp

Y. Chiba
National Research Institute for Earth Science and Disaster Resilience, Tsukuba, Japan
e-mail: y.chiba@bosai.go.jp

Md. A. Islam
Khulna University, Khulna, Bangladesh

V. Wijenayake et al. (eds.), *Linking Climate Change Adaptation, Disaster Risk Reduction, and Loss & Damage*,
https://doi.org/10.1007/978-981-99-8055-0_7

and damages (NELDs). This chapter highlights the importance of identifying appropriate NELD indicators and practices to help mitigate them. The findings suggest that community members and officials have different levels of understanding of NELDs, and practices like insurance may not be effective compared to preparedness planning. We suggest the localization of NELD indicators and identifying practices that can mitigate NELDs in a participatory manner. We suggest integrating NELD indicators into local post-disaster L&D assessment formats for quick assessment and use in recovery and long-term risk reduction planning. We emphasize the need for more information on the nature and extent of NELDs caused by climatic events to facilitate better mitigation. The study also identifies the need for addressing related institutional and technical challenges.

Keywords Loss and damage · Non-economic loss and damage · Indicators · Analytical hierarchy process · Multi-criteria decision-making · Adaptation planning

1 Introduction

Highlights

- Non-economic loss and damages (NELDs) need renewed attention by both the disaster risk reduction (DRR) and climate change adaptation (CCA) communities.
- Measuring/assessing NELDs requires identifying locally appropriate indicators in a consultative manner that can identify appropriate practices to mitigate NELDs.
- The NELD indicators need to be incorporated into post-disaster data collection formats used by local administration so that the information can be collected that will be useful for long-term risk reduction.

Loss and damages (L&Ds) accrued from disasters and climate change are at the centre of our strategies to mitigate the disaster and climate change threats. L&Ds provide us with a view of risks, underlying vulnerabilities and strengths and weaknesses of institutional and social systems. Hence, measuring L&Ds has been the priority of both the disaster risk reduction (DRR) communities and the climate change adaptation (CCA) communities. However, the way these two communities approached the issue of L&Ds differed.

The basic intent of measuring and or assessing L&Ds by DRR community has been to design immediate relief, response, and long-term recovery operations. This information has also helped to understand the progression of risks over time, to assess the efficacy of practices implemented over the period, and to an extent to assist future planning. On the other side, the CCA community looked at the L&Ds from the point of view of knowing the extent of L&Ds the climate change will cause beyond the possible adaptation practices already implemented or in the future. The topic of L&D has also been discussed among the climate change community focusing on the issue of additionality and causality of climate change, i.e., how much additional L&Ds can be accrued to climate change.

The United Nations Framework Convention on Climate Change (UNFCCC) defined L&D as "the actual and/or potential manifestation of impacts associated with climate change in developing countries that negatively affect human and natural systems" (UNFCCC 2012). There has been a wide opinion that the definition provided by UNFCCC doesn't satisfy the intent and aspirations of the parties to the Convention and that there is a need to refine this definition to provide clear operational guidance to the stakeholders (Boyd et al. 2017). To a large extent, the focus on L&D emerged internationally under UNFCCC negotiations focusing on liability and compensation and eventually moved on to focus on insurance and risk management (Vanhala and Hestbaek 2016).

Notwithstanding the differential perceptions and interpretations of L&D by the DRR and CCA community, understanding and mitigating losses and damages is of at most importance for guiding the practices implemented by both these communities. It is in this light this chapter discusses the L&Ds.

In general, L&Ds can be classified as economic loss and damages (ELDs) and non-economic loss and damages (NELDs). ELDs are defined as "The loss of resources, goods and services that are commonly traded in markets" (UNFCCC 2013). They also have been stated as those losses and damages that are "objectively verifiable monitory losses" (Fischer 2010). These can be understood as those losses and damages that have a market value or those that can be readily addressed by economic means. However, the ability to address by economic means or by those goods and services available from the market may not be equally applicable to both the L&Ds as inherently losses are those that cannot be brought back economically or otherwise. NELDs on the contrary can be understood as those losses and damages that do not have a market value or cannot be addressed by goods and services available from the marketplace. NELDs are defined as "those that are not commonly traded in markets" (UNFCCC 2013), and "subjective and non-verifiable losses" (Fischer, 2010). Major economic and non-economic losses and damages are listed in Table 1. It can be seen that the economic losses and damages are similar between sudden-onset and slow-onset climatic events while they differ in their NELDs.

In the subsequent section, we will understand that NELDs have been receiving limited attention in the work of the DRR and CCA communities. As NELDs have often not been measured, quantified, or assessed at the level they deserve, there is a dearth of information on the NELDs of past disasters including climatic events. This can lead to an underestimation of disaster and climate change vulnerabilities and risks and can lead to limited disaster recovery and climate change adaptation. To improve DRR and CCA, it is imperative that NELDs are measured, quantified, or assessed, and that the resulting information is used for risk assessment and management planning.

Table 1 Examples of economic and non-economic losses and damages

Climate events	*Economic L&Ds*	*Non-economic L&Ds*
Extreme weather/climatic events: e.g. *Typhoons, Storms, Floods, Cyclones,* etc.	• Damages to assets • Loss of income • Loss of production • Reduction in tourism revenue	• Lives lost • Human health • Displacement • Loss of cultural heritage
Slow onset events: e.g. *Sea level rise, Salinization, Drought,* etc	• Damages to assets • Loss of income • Loss of production • Reduction in tourism revenue	• Human health deterioration • Forced displacement • Uninhabitable territory • Damages to cultural heritages • Loss of indigenous knowledge • Loss of biodiversity and ecosystem

Source Prepared based on Chiba et al. (2017)

Prioritization of indicators of NELDs assumes importance in this context as not all NELD indicators may be applicable in all contexts and the context-specificity demands that they are identified for location-specific contexts (geographical, climate, socio-economic) in which decisions are made and appropriate practices are implemented. Keeping this need in view, this paper outlined the methodology to prioritize NELD indicators that can be implemented at the local level qualitatively using participatory techniques. It shows how the process of identifying NELD indicators can lead to the identification of practices that can connect to the objective of DRR and CCA set at the local level.

2 The Need for Focusing on NELDs

The authors focused the work presented in this chapter on NELDs for several reasons discussed here. Historically, economic losses and damages have been receiving much attention. On the contrary, the non-economic L&Ds have not been sufficiently reported in most post-disaster reports and databases (SwissRe 2013). A review of the L&D databases revealed that the national databases tend to be more centred on economic losses and damages than non-economic losses and damages (Table 2). The less emphasis on NELDs is understandable as economic losses and damages have traditionally been considered a priority for the recovery aftermath of a disaster, most of this recovery has been considered as having economic implications, and relief payments in terms of cash are much easier to implement especially the aftermath of major disasters. Hence, governments focused on economic aspects of losses and damages when the post-disaster impact measurements and assessments are conducted.

However, non-economic L&Ds can be significant and under certain contexts, they can be much more significant than economic losses and damages. They can be more important because the aspects covered by NELDs provide a strong basis for resilience, long-term and effective recovery, and general well-being including environmental well-being. The NELD elements provide the necessary foot-holding for the social fabric, form the basis for trust in institutions to function properly, contribute to dissipating the stress emanating from the disaster and climate change impacts, and help mitigate underlying vulnerabilities including those related to biodiversity and ecosystem services a large part of which is related to NELDs. Hence, without considering NELDs, the recovery

Table 2 Number of economic and non-economic L&D indicators covered by various disaster databases

Database	*The number of indicators reported*	
	Economic	*Non-economic*
EM-DAT	1	5
Japan (Database covering natural disasters during 2003–2011)	10	5
Bangladesh (database covering floods, cyclones and landslides)	8	3

Source Authors

process can be long-drawn, incomplete, and unsustainable. If unaddressed, the diminishing NELD components can increasingly expose human and environmental systems to future climate impacts.

NELDs are also currently far less understood than ELDs, as there are not sufficient assessment frameworks for assessing NELDs both within the DRR and CCA domains. NELDs have also not been well considered in climatic and non-climatic risk assessments and including in designing insurance and compensation mechanisms (Hoffmaister and Stabinsky 2012). NELDs such as mental disorders, psychological and social impacts of people displaced, damages to social and cultural capitals, and damages to biodiversity/ecosystem are seldom measured aftermath of disasters and climatic events.

Our current CCA and DRR institutions are also not well developed to adopt and value NELD in decision-making largely because of economic-driven thinking, cost-driven decision-making and lack of progress in 'valuing the invaluable'. To a certain extent, providing decision-relevant information can be a challenge more for NELDs than for ELDs. Often the skills are not sufficiently developed to better present NELDs in a fashion that decision-makers can readily use. On the contrary, social systems tend to value NELDs much higher than one would think, much higher than how formal institutions value the NELDs. As we would see in the results section, the way the society members perceive NELD indicators can be different from government officials. This brings the aspect of differential perceptions of NELDs by various stakeholders and signifies the need to have a consultative dialogue and engagement among the stakeholders to achieve a consensus and common understanding on NELDs.

The focus on social and natural elements of vulnerability and adaptive capacity has been more recent as historically the focus has been on post-disaster relief focusing heavily on providing economic relief. Since NELDs are far less understood, less measured, less quantified, and less assessed, our risk and vulnerability assessments, both within the works of the DRR and CCA community, can be stated as incomplete. As a result, we may have largely been underestimating the disaster and climate change impacts, vulnerabilities and risk projections.

This will have the undesirable consequence of our DRR plans and CCA plans falling short of expectations as they can fail to mitigate a large proportion of vulnerabilities that are related to NELDs leading to the long-term perpetuation of vulnerabilities, risks, losses and damages and can lead to the phenomenon of 'hidden vulnerabilities' only to be surfaced aftermath of a catastrophic event.

Underemphasizing NELDs can also lead to the identification of inappropriate DRR and CCA strategies as well. Since the understanding of underlying conditions of vulnerabilities plays an important role in identifying solutions, any solutions identified only based on economic impacts will not mitigate the entire spectrum of impacts experienced by disasters and climate change. This can lead to insufficient recovery and limited progress in DRR and CCA.

2.1 Better Measuring NELDs and Linking with Risk Reduction Solutions

For making any progress in understanding and mitigating NELDs, the NELDs must be measured aftermath of the disasters and climatic events so that this information can be used for risk assessments, risk management planning, and better recovery from disasters and climatic events. With this information and understanding, the future projections can also be improved subsequently which will help us to prepare well for the future.

There are several L&D assessment methodologies available (Table 3). These methodologies vary in their technical rigour, the skill requirements of users, and the nature of the information output they produce. Though the users of these methods found certain advantages, they suffer from the limitation that they are largely not accessible for most local officials that engage in measuring/assessing L&Ds. Most assessment methods are focused on economic L&Ds, they are complex in nature and hence not accessible to most stakeholders. Most of them also take a 'steep learning curve' as actors engaged in DRR (and to an extent in CCA) are not well-versed with the non-economic valuations of L&Ds. Measurement of losses and damages aftermath of disasters is often carried out by officials on the ground and they can benefit from the use of simplified methodologies.

Since NELDs are location-specific reflecting the local environmental and socio-cultural contexts, NELDs must be assessed using a locally identified set of indicators that better reflect the local contexts. The process of identifying indicators also needs to contribute to the identification of effective practices. Otherwise, the lack of strong linkage between indicators and practices could lead to erroneous DRR and CCA outcomes.

Efforts for prioritizing NELD indicators and practices need to essentially recognize not only the fact that they are location-specific but also need to satisfy the consultative requirements in identifying them as multiple stakeholders engaged in DRR and CCA may value NELDs differently and they may have multiple criteria in selecting indicators and practices. Due to this subjective nature, the identification of NELDs must engage methodologies that can accommodate multiple criteria of stakeholders in a consultative setup. For this reason, multi-criteria assessment (MCA) methodologies suit well for identifying NELD indicators.

MCA methodologies aid in selecting the 'best' alternative from the number of feasible choice alternatives under the presence of many criteria and diverse criterion priorities. Several multi-criteria methods are available that can be readily used in the context of NELD and prominent ones being cost–benefit analysis, cost-effective analysis and analytical hierarchy process (AHP) (Prabhakar and Krishna 2013). While cost–benefit and cost-effectiveness analysis tend to rely heavily on quantified costs and benefits, AHP provides the flexibility of using qualitative elements. For this reason, the authors used the AHP to prioritize NELD indicators and practices. AHP helps in structuring a multi-dimensional problem into a hierarchical tree with criteria and alternatives/practices. Developed by Saaty (1990), the method has received wide attention and application in wider decision-making fields. It has been regarded as one of the easiest-to-use MCA techniques, easy to interpret results, useful in the evaluation of projects and policies, compatible with other quantitative methods, and effectively links the indicators and alternatives/practices

Table 3 Methodologies for assessing L&D

Quantitative or qualitative	*Examples of approaches*	*Overview*	*Hazard type*
DRR (pre-disaster)			
Quantitative	A comprehensive approach to probabilistic risk assessment	Probabilistic risk assessment based on the GIS platform	Earthquakes; tsunamis; hurricanes; floods; landslides; volcanoes
	Catastrophe simulation model of the IIASA	Monte Carlo simulation of disaster risks which examines the fiscal and economic risk	Floods; hurricanes; weather and climate-related hazards; earthquakes
Qualitative	Community-based disaster risk management (CBDRM)	Application of measures in risk analysis, disaster prevention and mitigation and disaster preparedness by local actors	Droughts; heatwaves; floods; hurricanes; earthquakes; volcanoes
	Vulnerability and capacity assessment (VCA)	The basic process is used to identify the strengths and weaknesses of households, communities, and institutions to support decisions made in the development of mitigation programmes	Droughts; floods; earthquakes
DRR (post-disaster)			
Quantitative	Economic Commission for Latin America and the Caribbean	A handbook that describes the methods required to assess the social, economic and environmental effects of disasters	Floods; hurricanes; weather and climate-related hazards; earthquakes
	Emergency Management Australia (EMA)	Guidelines that explain the process of loss assessment, through the steps required to carry out an economic assessment of disaster losses	Floods; hurricanes; weather and climate-related hazards; earthquakes
Qualitative	CBDRM	Same as above	Same as above
	VCA	Same as above	Same as above
Climate change adaptation			

(continued)

Table 3 (continued)

Quantitative or qualitative	*Examples of approaches*	*Overview*	*Hazard type*
Quantitative	Integrated impact assessment models	Model for the dynamics of carbon accumulation in the atmosphere and their influence on the economy	No specific hazard focus
	Country environmental analysis (CEA), Strategic environmental assessment (SEA)	Analytical tools on the prioritization of environmental issues in terms of their effect on economic development and poverty	Droughts; land degradation; floods; hurricanes
Qualitative	UKCCRA	Same as above	Same as above

Source Authors based on Surminski et al. (2012) andUNFCCC (2013)

(Prabhakar and Krishna 2013; Saaty 1990). AHP also helps capture both subjective and objective evaluation measures and alternatives/practices, the pair-wise comparison is easy to understand, and group decision is supported through consensus by calculating the geometric mean of the individual pair-wise comparisons and reduces bias in decision-making. AHP is also known to offer effective means in situations of uncertainty and risk through derivation of scale where measures do not exist.

3 Methodology

The methodology selection in this research was driven by the factors such as the requirement of consultation with stakeholders as multiple stakeholders have multiple criteria in identifying indicators and practices and how they value NELDs, being able to adopt the method in varied contexts, to be able to provide an output that is easy to use and interpret by the decision-makers engaged in DRR and CCA and be able to link the NELD indicators with practices to mitigate NELDs.

The overall methodological flow for identifying indicators and practices for NELDs is shown in Fig. 1. The sequence of steps for identifying and prioritizing NELD indicators and practices can be broadly divided into (1) Identification of location for carrying out the exercise, (2) Identification of a broad set of indicators and practices that needs to be further vetted, verified, and narrowed down to the location-specific conditions, and (3) AHP exercise further helps in narrowing down and prioritizing among the narrow set of indicators and practices.

1. Identification of indicators and practices for NELD: This was primarily accomplished by conducting extensive literature reviews and tabulation of indicators and practices for mitigating NELDs (Prabhakar and Krishna 2013; Chiba et al. 2017, 2018). As the literature targeting NELDs purely focused on climate change is very limited, a broad scope of literature was reviewed including that of climatic events. 31 indicators in 10 thematic areas and 25 practices were identified. This tabulated list of NELD indicators and practices formed a basis for further consultations with the experts.
2. Expert consultation meetings were organized in both the case study countries of Bangladesh and Japan to further narrow down these indicators and practices to the location-specific conditions. The consultations were attended by CCA specialists, the water resources department, focal personnel of disaster risk management, local NGOs and elected members and community members of Koyra,

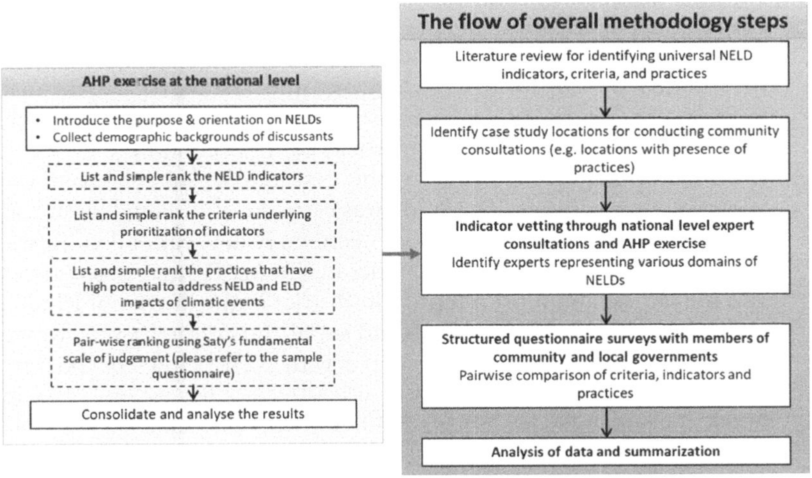

Fig. 1 Methodological flow followed in the research for identifying and prioritizing indicators and practices for NELD

Bangladesh and Nachikatsuura, Japan. The process for narrowing down consisted of extensive discussion on each indicator and practice, and reliance on the knowledge and experience of the experts who attended the consultation. Simple ranking and rating techniques were employed to narrow down the list after the discussions. AHP exercise was organized with the same stakeholders who participated in narrowing down the NELD indicator and practices to local contexts. The AHP exercise consisted of introducing the purpose of the AHP exercise, agreeing on the objective of the AHP outcome (in the case of Bangladesh the objective identified was 'to reduce NELD due to extreme cyclones' and in Japan, the objective identified was 'to reduce the NELD due to extreme typhoons'. After agreeing upon the objective, the pair-wise comparison of criteria was carried out using Saaty's fundamental scale of judgement which has a scale of 1–9 (Saaty 1990). Here, the criteria indicate the logic and value judgement assigned by the respondent in choosing a particular indicator or practice against another indicator or practice. The AHP exercises were carried out using the Super Decisions software that provides the result of the AHP exercise systematically and enables to obtain a consensus input to the AHP exercise through discussion among the participants.

3. Identification of location: Location identification was primarily determined by the local knowledge of the research partners that satisfies the criteria of (a) occurrence of a major climatic event, and (b) providing a contrasting socio-cultural condition to see if the methodology employed by the research is suitable to these contrasting conditions. The study was carried out in Bangladesh and Japan which have contrasting socio-economic and cultural contexts. In Bangladesh, the Koyra sub-district in Khulna District was chosen. It is the largest and most remote sub-district in the district of Khulna and enjoys rich natural resources and social and cultural assets. Koyra has a population of 193,931 people with 32% below the poverty (Chiba et al. 2018). The region was affected by cyclone Aila in 2009. The cyclone has affected 300,000 people, killed 57 people and displaced several hundred. 7392 ha of the cropped area was destroyed due to the cyclone. A large number of people experienced a shortage of food, water, shelter, medication, and sanitation facilities. In Japan, the study was carried out in the Nachikatsura town of Wakayama prefecture which was severely affected by Typhoon

No 12 in 2017. The town is located in the southeast part of the Wakayama prefecture with a total population of 15,946 with <2% poverty. The typhoon affected 2410 households, killed 29 people and 14,458 people were evacuated (Chiba and Prabhakar 2017a, b).

4. Structured questionnaire surveys: Structured questionnaire surveys were conducted with the participation of community members and local government officials in Kyora, Bangladesh and Nichikatsura, Japan. These questionnaires utilized the expert consultations conducted at the national level in both the countries where the criteria, indicators, and practices were further narrowed down to the local context. The purpose of conducting questionnaire surveys with community members and local governments was to further localize these indicators and practices to the local city contexts. Stratified random sampling was conducted to ensure the participation of gender, age, and economic categories of communities as proportionately represented in the latest censuses data. 237 questionnaires were returned by the community members in Bangladesh, and 175 questionnaires were returned in Japan. The participation of local government officials was also important aspect of the study to understand the differences in perspectives between the local community members and local government officials. Consequently, 32 questionnaires were returned by the local government officials in Bangladesh and 22 questionnaires were returned by the local government officials in Japan. These results were entered into an excel sheet, and AHP analysis was carried out for the responses received. In this chapter, only the qualitative results are presented from the questionnaire surveys, i.e., the type of criteria, indicators, and practices prioritized by the respondents. For quantitative results, the reader is advised to refer to the research report entitled 'Addressing Non-economic Losses and Damages Associated with Climate Change: Learning from the Recent Past Extreme Climatic Events for Future Planning'.

4 Results and Discussion

Overall, organizing an AHP exercise with multiple stakeholders provides a rich experience of deep discussions among the participants. It helps to deepen the understanding of issues and effectively helps link issues with solutions. Though time-consuming, the efforts put into organizing such consultations are well satisfied by the valuable insights and learnings from the experience. Specific results from both country locations are discussed below.

4.1 *Koyra, Bangladesh*

Figure 2 shows the overall results of NELD-related elements prioritized by community members in Koyra. 'Compliance with societal value' was a principal criterion for decision-making. The community members gave more emphasis on water and sanitation indicators including 'inaccessible sanitation' and 'waterborne diseases, and a health-related indicator of 'mental health disorder'. They in turn determined that the region-specific 'DRR policy and planning' was the most effective risk reduction practice for NELDs.

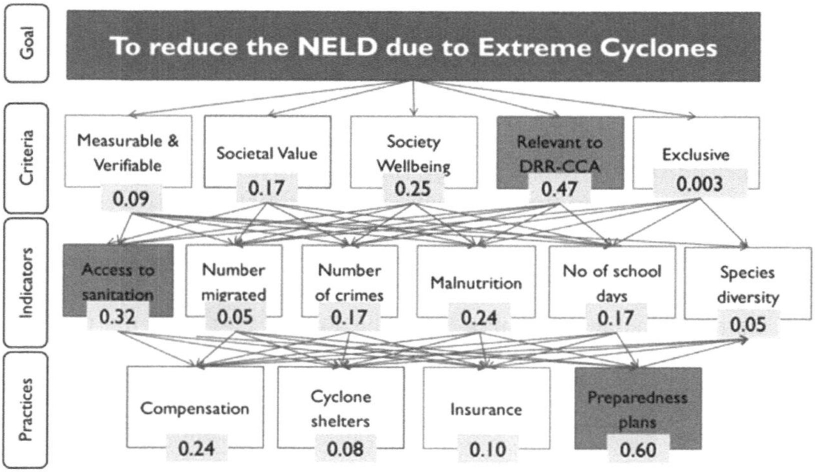

Fig. 2 Analytical hierarchy process (AHP) results from Bangladesh

The results have shown a difference in perspectives between government officials and community members. As shown in Table 4, 'Relevance to DRR/CCA policy and planning' was considered the most important criterion for identifying the appropriate indicators and practices. These indicators and practices were identified from the questionnaire surveys conducted with the community members and government officials. The government officers also identified 'waterborne diseases', 'inaccessible sanitation', and 'school discontinued' as important indicators for assessing the effectiveness of practices and that 'DRR policy and planning as the most effective practice to mitigate NELDs.

Figure 2 shows the results from the expert consultation workshop conducted at the national level where AHP exercise was organized. This exercise provided a deeper insight into how various experts think about NELDs. The opportunity to bring experts from different domains together in a systematic setup of AHP enabled the study team to understand the deep differences and be able to negotiate for a common understanding among them. Such an effort is necessary for all the NELD issues going forward as it will lead to much more ownership in the outcomes and greater uptake of practices that emerge from these negotiations.

The result shows that the important criteria for identifying NELD indicators are their relevance to DRR and CCA. The indicators also need to satisfy the criteria that they need to be relevant to the overall societal

Table 4 List of NELD-related elements prioritized by community members and government officials in Koyra, Bangladesh

Rank	*Criteria*	*Indicators*	*Practices*
Community members			
1	Compliance with societal value	Inaccessible sanitation	DRR policy and planning
2		Waterborne diseases	
3		Mental health disorder	
Government officials			
1	Relevance to DRR/CCA policy and planning	Waterborne diseases	DRR policy and planning
2		Inaccessible sanitation	
3		Schools discontinued	

Source Authors

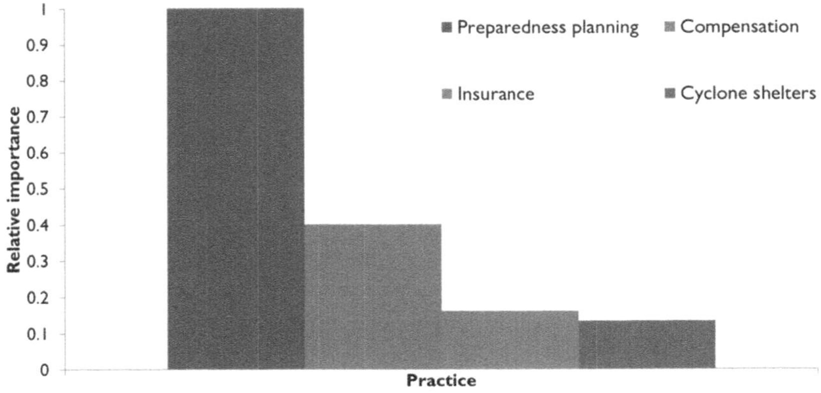

Fig. 3 Efficacy of practices for addressing NELDs in Bangladesh

well-being, not just economic well-being. Interestingly, the measurability and verifiability of indicators were not found to be as important as a researcher or a development professional would consider them to be. The participants also opined that the indicators need not be exclusive to NELDs themselves and that the societal value of NELDs may not have to be an important determinant in their identification. Among the practices, preparedness planning received the highest importance followed by compensation, insurance and cyclone shelters. The reason behind the higher preference for preparedness planning could be that evacuation of community members before a projected extreme event can mitigate the loss of human life, instils cooperation among the community members, and helps prioritize the most vulnerable members of the community. AHP provides relative importance values for the practices as well, and this relative importance can help us to find a combination of practices to replace the equally effective practice (Fig. 3). It can be seen that in absence of preparedness planning, one would need compensation, insurance, and cyclone shelters combined to achieve the equivalent outcome of preparedness planning.

4.2 *Nachikatsuura, Japan*

Figure 4 presents the overall results of NELD-related elements prioritized by experts in the national level workshop organized in Tokyo, Japan.

Similar to Bangladesh's case, 'compliance with societal value' appears to be the dominant criteria for decision-making, and community members prioritized local governance indicators such as 'less collaboration of local government with local communities, 'less participation of the community in decision-making, and 'mental health disorder' as important indicators for identifying effective practices for mitigating NELDs. The community members prioritized the 'emergency shelters' for securing safe locations for local communities to be the most effective risk reduction practice for NELDs. The focus on emergency shelters indicates the short-term nature of the NELD impacts, which was not the case in Bangladesh where long-term impacts of NELDs appeared to have dominated the decision-making.

Similar to Bangladesh, the Japanese experts also opined that the NELD indicators need to be relevant to both DRR and CCA. Unlike in Bangladesh, the Japanese experts considered societal values as an important criterion. Participants from both countries considered the measurability and verifiability of indicators as the least important among the prioritized criteria. Among the indicators, human life received the highest importance followed by post-traumatic stress disorder (PTSD),

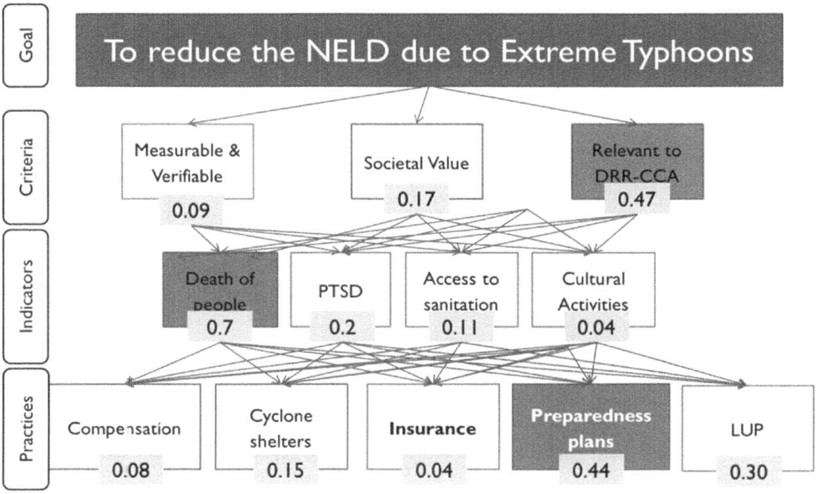

Fig. 4 Analytical hierarchy process (AHP) results from Japan

access to sanitation and cultural activities. Among the practices, preparedness planning received priority over other practices as seen in Bangladesh. The relative importance of practices in Japan is presented in Fig. 5. In the case of Japan, preparedness planning can be replaced by a combination of land use planning and cyclone shelters unlike in Bangladesh where more practices had to be combined to equal the preparedness planning.

Questionnaire surveys were conducted at the community level using structured questionnaires consisting of pair-wise comparison of criteria, indicators, and practices for AHP analysis. The results of the questionnaire survey are presented in Table 5. 'Compliance with societal value' was considered most important by the community members and local government officials. The indicators of 'mental health disorder', 'less collaboration of local government with local communities, and 'chronic diseases' were the three most prioritized indicators, and, in contrast to the community perspective, 'DRR policy and planning' was the most effective risk reduction practice for NELDs indicating the long-term view taken by the government officials.

Comparing results between Bangladesh and Japan may provide some insights into the association between the developmental state of the country and the opinions of respondents in terms of NELD indicators and practices. The results indicate that both Bangladesh and Japan placed high importance on addressing issues with mental health disorders (Table 6). Different characteristics of these countries in terms of NELDs could also

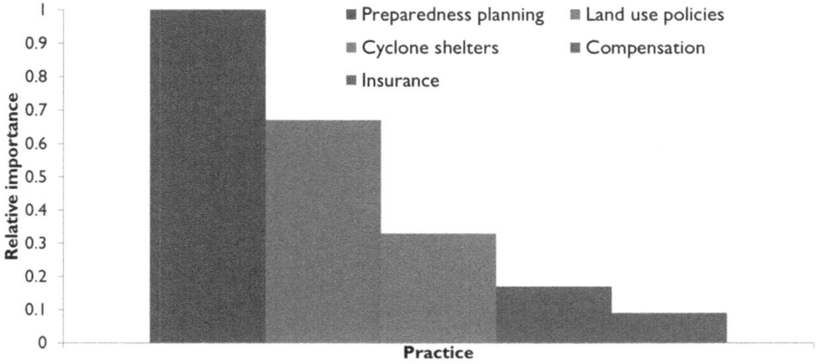

Fig. 5 Efficacy of practices for mitigating NELDs in Japan

Table 5 List of NELD-related elements prioritized by community members and government officials in Nachikatsuura, Japan

Rank	*Criteria*	*Indicators*	*Practices*
Community members			
1	Compliance with societal value	Less collaboration between the local government and the community	Emergency shelters
2		Less community participation in decision-making	
3		Mental health disorder	
Government officials			
1	Compliance with societal value	Mental health disorder	DRR policy and planning
2		Less collaboration between the local government and the community	
3		Chronic diseases	

Source Authors

be found. In the case of Bangladesh, issues such as inaccessible sanitation, and waterborne diseases took precedence in discussions with the community members and government officials. This is in line with the significant challenge that especially developing countries are facing after cyclones due to the breakdown of water and sanitation systems and the dearth of safe drinking water (Haque et al. 2012). Conversely, Japan's case has highlighted challenges in local risk governance raising the need for close coordination and communication with community associations, voluntary organizations, and volunteer groups, and for establishing communication channels to seek opinions and consensus-building with communities.

Table 6 Comparison of the top three NELDs identified individually by the community and government officials in Bangladesh and Japan

NELDs		*Bangladesh: Koyra*		*Japan: Nachikatsuura*	
Impact area	*Indicators*	*Community*	*Gov't*	*Community*	*Gov't*
Health	Mental health disorder	•		•	•
	Chronic diseases				•
Water & Sanitation	Inaccessible sanitation	•	•		
	Waterborne diseases	•	•		
Education	Schools discontinued		•		
Local governance	Less collaboration			•	•
	Less participation			•	

Note Gov't = Local government officials, • = prioritized
Source Authors

5 Conclusion

This study intended to understand NELDs caused by recent past major climate-related disasters and developed a methodological framework to identify and prioritize NELDs and important practices to mitigate NELDs. The cases of Bangladesh and Japan were presented. The AHP was applied to identify and prioritize key NELDs caused by climate-related disasters and to find important risk reduction practices that could mitigate NELDs.

For Koyra, from the community perspective, the study identified 'compliance with societal value' as an important criterion, 'inaccessible sanitation', 'waterborne diseases' and 'mental health disorder' as important indicators, and 'DRR policy and planning as an important risk reduction practice. On the other hand, for Nachikatsuura, the study specified 'compliance with societal value', 'less collaboration of local government with local communities', 'less participation of the community in decision-making', 'mental health disorder', and 'emergency shelter' as important NELD elements. The results indicate that both Bangladesh and Japan placed high importance on addressing issues with mental health disorders. Bangladesh's case has identified serious issues of inaccessible sanitation and waterborne diseases, while Japan's case has highlighted challenges in local risk governance in terms of communication between local government and communities.

There is a need that these prioritized NELDs to be incorporated into the post-disaster needs assessments (PDNAs), and post-disaster impact assessments conducted in both countries. The local governments in both countries have damage assessment formats that are filled aftermath of disasters and the prioritized NELDs may be incorporated into these formats for regular data collection and use for risk reduction decision-making. The collected information will be part of a database that helps in understanding the progression of NELDs over a period giving a deeper understanding of how NELDs accumulate. Incorporating these NELDs into vulnerability and risk assessments for the future will help prepare governments and communities and help mitigate NELDs effectively. Existing challenges such as a lack of understanding of NELDs and skills to measure and or assess their needs to be addressed through capacity-building activities targeting the specific representatives engaged in vulnerability and risk assessments, risk management planning, and implementation of risk reduction activities on the ground.

Acknowledgements We gratefully acknowledge several researchers, government officials, non-governmental organizations and community members who participated in this research directly and indirectly by providing valuable time, experiences, and expertise, including being part of the workshops, consultations and surveys conducted in this research. The authors are grateful that part of this work research was supported by the Environment Research and Technology Development Fund (2-1801) of the Environmental Restoration and Conservation Agency of Japan.

Bibliography

Boyd, Emily, Rachel James, Richard Jones, Hannah R. Young, and Friederike E. L. Otto. 2017. "Typologies of loss and damage and associated actions." *Nature Climate Change* 7: 723–729.

Chiba, Y., Rajib Shaw, and Sivapuram Venkata Rama Krishna Prabhakar. 2017. "Climate change-related non-economic loss and damage in Bangladesh and Japan." *International Journal of Climate Change Strategies and Management* 9 (2): 166–183.

Chiba, Yohei, and Sivapuram Venkata Rama Krishna Prabhakar. 2017a. *Priority practices for addressing non-economic loss and damage caused by Typhoons in Japan: Case study of Nachikatsuura Town*. IGES Research Report, Hayama, Japan: Institute for Global Environmental Strategies, 22.

Chiba, Yohei, and S.V.R.K. Prabhakar. 2017b. *Addressing non-economic losses and damages associated with climate change: Learning from the recent past extreme climatic events for future planning*. Kobe, Japan: Asia-Pacific Network for Global Change Research (APN) and Institute for Global Environmental Strategies (IGES), 142.

Chiba, Yohe, Sivapuram Venkata Rama Krishna Prabhakar, M. A. Islam, and M. A. Akber. 2018. "Priority practices for addressing non-economic loss and damages caused by cyclones in Bangladesh: Case study of Koyra." *International Journal of Disaster Resilience in the Built Environment* 9 (4/5): 333–347.

Fischer, J.M. 2010. *Understanding remedies*. Dayton, OH: LexisNexis.

Haque, U., M. Hashizume, K.N. Kolivras, H.J. Overgaard, B. Das, and T. Yamamoto. 2012. Reduced death rates from cyclones in Bangladesh: What more needs to be done? *Bulletin of the World Health Organization* 90 (2): 150–156.

Hoffmaister, J.P., and D. Stabinsky. 2012. "Loss and damage: Some key issues and considerations for SIDS expert meeting." *SIDS Expert Meeting*. Penang: Third World Network.

Prabhakar, Sivapuram Venkata Rama Krishna. 2013. *Adaptation decision making frameworks and tools: Multi-criteria decision making tools for prioritizing adaptation actions at community level*. Hayama, Japan: Institute for Global Environmental Strategies.

Saaty, T. L. 1990. "How to make a decision: The analytic hierarchy process." *European Journal of Operational Research* 48 (1): 9–26.

Surminski, S., A. Lopez, J. Birkmann, and T. Welle. 2012. *Current knowledge on relevant methodologies and data requirements as well as lessons learned and gaps identified at different levels, in assessing the risk of loss and damage associated with the adverse effects of climate change*. Technical Paper, Bonn, Germany: UNFCCC, 45.

SwissRe. 2013. *Natural catastrophes and man-made disasters in 2012: A year of extreme weather*. Zurich, Swtizerland: SwissRe.

UNFCCC. 2012. *A literature review on the topics in the context of thematic area 2 of the work programme on loss and damage: A range of approaches to address loss and damage associated with the adverse effects of climate change*. Bonn, Germany: UNFCCC.

UNFCCC. 2013. *Non-economic losses in the context of the work programme on loss and damage*. Bonn, Germany: UNFCCC.

Vanhala, Lisa, and Cecilie Hestbaek. 2016. "Framing climate change loss and damage in UNFCCC negotiations." *Global Environmental Politics* 16 (4): 111–129.

CHAPTER 8

Climate Change-Induced Migration in South Asia

Amit Ranjan

Abstract This chapter discusses how climate change is becoming a significant driver of population migration, as one report estimates, with over one billion people are expected to be displaced due to natural disasters and ecological events by 2050. South Asia is one of the regions facing the most significant number of ecological threats. Climate change will also lead to economic and non-economic damages, with losses including agricultural production, property and infrastructure damage, and the loss of cultural traditions, indigenous knowledge, biodiversity, and ecosystem services. The lack of resilience in many countries in South Asia may cause to worsening food insecurity, civil unrest, and mass displacement.

Keywords South Asia · Migration · Climate change · Resilience

A. Ranjan (✉)
Institute of South Asian Studies, National University of Singapore, Singapore, Singapore
e-mail: amitranjan.jnu@gmail.com

V. Wijenayake et al. (eds.), *Linking Climate Change Adaptation, Disaster Risk Reduction, and Loss & Damage*,
https://doi.org/10.1007/978-981-99-8055-0_8

1 Introduction

The migration of people from one place to another has been happening since ancient times. People migrate for various reasons, one of which is human-influenced climate change. In modern times, at many places in the world, climate change and its wide impact are turning into chief drivers of population migration from their place of living. Such migration is mainly to protect life or to earn a livelihood.[1] In the coming years, the number of people migrating due to climate change will likely to increase. According to the 2020, Ecological Threat Report released by an international think tank, the Institute for Economics & Peace, more than one billion people will be displaced due to climate change and natural disasters. Key results from the report's findings are[2]:

1. Around 19 countries with the highest number of ecological threats are among the world's 40 least peaceful countries including Afghanistan, Syria, Iraq, Chad, India and Pakistan.
2. Over one billion people live in 31 countries where the country's resilience is unlikely to sufficiently withstand the impact of ecological events by 2050, contributing to mass population displacement.
3. Sub-Saharan Africa, South Asia, the Middle East, and North Africa are the regions facing the largest number of ecological threats.
4. By 2040, a total of 5.4 billion people—more than half of the world's projected population—will live in the 59 countries experiencing high or extreme water stress, including India and China.
5. About 3.5 billion people could suffer from food insecurity by 2050; which is an increase of 1.5 billion people from today.

[1] See, Ranjan, Amit "Climate Change, Environmental Migration and Population Displacement" in Amit Ranjan, Rajesh Kharat and Pallavi Deka (edited) *Environment, Climate Change and Migration in South Asia*. Routledge: New Delhi, pp. 12–30.

[2] Institute for Economics & Peace (2020) "Over One Billion People at Threat of Being Displaced by 2050 Due to Environmental Change, Conflict and Civil Unrest" https://www.economicsandpeace.org/wp-content/uploads/2020/09/Ecological-Threat-Register-Press-Release-27.08-FINAL.pdf, p. 1. Accessed on 11 January 2023.

6. The lack of resilience in countries covered in the ETR will lead to worsening food insecurity and competition over resources, increasing civil unrest and mass displacement, exposing developed countries to increased influxes of refugees.

Moreover, global climate change will add to many other forms of economic and other non-economic damage. Intergovernmental Panel on Climate Change (IPCC) and many other institutional reports, policy papers, and media reports have repeatedly demonstrated the serious impact of climate change in South Asia. The social and financial impacts of climate change that human beings cannot avoid now are termed "loss and damage". Economic damage from climate change includes financial losses, such as loss of agricultural production due to heatwaves in India, affecting several farmers. The loss of property and infrastructure due to floods in many South Asian countries.[3] Non-economic loss and damage include loss of cultural traditions, indigenous knowledge, biodiversity, and ecosystem services.[4] A detailed definition of Loss and Damage was adopted at the 21st Conference of Parties meeting in Paris on 12 December 2015. Article 8 of the agreement says the following on Loss and Damage[5]:

1. Parties recognize the importance of averting, minimizing, and addressing loss and damage associated with the adverse effects of climate change, including extreme weather events and slow onset events, and the role of sustainable development in reducing the risk of loss and damage.

[3] Kumari, Shalinee (2022, 13 July) "What is Loss and Damage, and What Does It Mean for South Asia?" *The Third Pole*. https://www.thethirdpole.net/en/climate/what-is-loss-and-damage-south-asia/. Accessed on 19 September 2022.

[4] Ibid.

[5] "The Paris Outcome on Loss and Damage" https://unfccc.int/files/adaptation/groups_committees/loss_and_damage_executive_committee/application/pdf/ref_8_decision_xcp.21.pdf. Accessed on 19 June 2021.

2. The Warsaw International Mechanism[6] for Loss and Damage associated with Climate Change Impacts shall be subject to the authority and guidance of the Conference of the Parties serving as the meeting of the Parties to the Paris Agreement and may be enhanced and strengthened, as determined by the Conference of the Parties serving as the meeting of the Parties to the Paris Agreement.
3. Parties should enhance understanding, action, and support, including through the Warsaw International Mechanism, as appropriate, on a cooperative and facilitative basis with respect to loss and damage associated with the adverse effects of climate change.
4. Accordingly, areas of cooperation and facilitation to enhance understanding, action, and support may include: (a) Early warning systems; (b) Emergency preparedness; (c) Slow onset events; (d) Events that may involve irreversible and permanent loss and damage; (e) Comprehensive risk assessment and management; (f) Risk insurance facilities, climate risk pooling and other insurance solutions; (g) Non-economic losses; and (h) Resilience of communities, livelihoods, and ecosystems.
5. The Warsaw International Mechanism shall collaborate with existing bodies and expert groups under the Agreement, as well as relevant organizations and expert bodies outside the Agreement.

[6] At COP 19 meeting at Warsaw in Poland in November 2013, Warsaw International Mechanism for Loss and Damage associated with Climate Change was established. Its objective was to address loss and damage associated with impacts of climate change, including extreme events and slow onset events, in developing countries that are vulnerable to the adverse effects of climate change. The implementation was to be guided by the Executive Committee under the guidance of the COP. The Executive Committee has 20 members. 10 members are from annex I countries and 10 from non-annex I. Annex I consists of developed countries while non-annex I includes developing countries. The decision was to be taken by consensus. See United Nations Climate Change "Warsaw International Mechanism for Loss and Damage associated with Climate Change Impacts (WIM)". https://unfccc.int/topics/adaptation-and-resilience/workstreams/loss-and-damage/warsaw-international-mechanism#:~:text=The%20COP%20established%20the%20Warsaw,that%20are%20particularly%20vulnerable%20to. Accessed on 18 December 2022. United Nations Climate Change "Membership-L&D ExCom". https://unfccc.int/process/bodies/constituted-bodies/wim-excom/members. Accessed on 18 December 2022.

Further paragraphs 48–52 (Loss and Damage) says:

> 48. Decides on the continuation of the Warsaw International Mechanism for Loss and Damage associated with Climate Change Impacts, following the review in 2016;
>
> 49. Requests the Executive Committee of the Warsaw International Mechanism to establish a clearinghouse for risk transfer that serves as a repository for information on insurance and risk transfer, in order to facilitate the efforts of Parties to develop and implement comprehensive risk management strategies;
>
> 50. Also requests the Executive Committee of the Warsaw International Mechanism to establish, according to its procedures and mandate, a task force to complement, draw upon the work of and involve, as appropriate, existing bodies and expert groups under the Convention including the Adaptation Committee and the Least Developed Countries Expert Group, as well as relevant organizations and expert bodies outside the Convention, to develop recommendations for integrated approaches to avert, minimize, and address displacement related to the adverse impacts of climate change;
>
> 51. Further requests the Executive Committee of the Warsaw International Mechanism to initiate its work, at its next meeting, to operationalize the provisions referred to in paragraphs 49 and 50 above, and to report on progress thereon in its annual report;
>
> 52. Agrees that Article 8 of the Agreement does not involve or provide a basis for any liability or compensation.

In November 2022, at Sharm el-Sheikh in Egypt, the United Nations Climate Change Conference COP27 achieved a breakthrough when the parties agreed to provide compensation for Loss and Damage. For the first time in the official agenda, COP27 added the creation of a specific fund for Loss and Damage. The member countries also agreed to establish a "transitional committee" to recommend how to operationalize both the new funding arrangements and the fund at COP28 next year. The first meeting of the transitional committee will take place in March 2023. At Sharm el-Sheikh, following were also announced[7]:

[7] United Nations Climate Change "COP27 Reaches Breakthrough Agreement on New 'Loss and Damage' Fund for Vulnerable Countries" 20 November 2022. https://unfccc.int/news/cop27-reaches-breakthrough-agreement-on-new-loss-and-damage-fund-for-vulnerable-countries. Accessed on 18 January 2023.

- Countries launched 25 new collaborative actions in five key areas: power, road transport, steel, hydrogen and agriculture.
- UN Secretary-General António Guterres announced a US$3.1 billion plan to ensure everyone on the planet is protected by early warning systems within the next five years.
- The UN Secretary-General's High-Level Expert Group on Net-Zero Commitments published a report at COP27, serving as a how-to guide to ensure credible, accountable net-zero pledges by industry, financial institutions, cities, and regions.
- The G7 and the V20 ('the Vulnerable Twenty') launched the Global Shield against Climate Risks, with new commitments of over US$200 million as initial funding. Implementation is to start immediately.
- Announcing a total of US$105.6 million in new funding, Denmark, Finland, Germany, Ireland, Slovenia, Sweden, Switzerland, and the Walloon Region of Belgium, stressed the need for even more support for the Global Environment Facility funds targeting the immediate climate adaptation needs of low-lying and low-income states.
- The new Indonesia Just Energy Transition Partnership, announced at the G20 Summit held in parallel with COP27, will mobilize US$20 billion over the next three to five years to accelerate a just energy transition.
- Important progress was made on forest protection with the launch of the Forest and Climate Leaders' Partnership, which aims to unite action by governments, businesses and community leaders to halt forest loss and land degradation by 2030.

The first part of this chapter examines the theory of climate-induced human migration. The second part discusses South Asia's climate situation and then examines its impact on the respective countries. This chapter argues that the number of internally displaced populations due to climate change will likely to increase further in South Asia. South Asian countries adopt policies to deal with climate change; however, proper execution and implementation of those policies remain a problem in all South Asian countries. Some South Asian countries suffer from a rise in the emission of Green House Gases (GHG) despite being low at the table in the global emitters list.

2 Climate Change-Induced Migration—Concept, Pattern, and Terms

There has been a considerable increase in literature on climate migration since the 2000s. However, there is an insufficient focus on how the study of climate migration engages with theory. But a consensus is growing on the need to "embed climate mobility research in migration theory".[8] The Foresight report that incorporates a number of the elements of other theories, including neoclassical theory, migrant selectivity, world systems, and push–pull, considers climate change as an indirect driver of migration which operates on the direct drivers such as the pre-existing economic, demographic, social, political, and environmental conditions.[9] Even earlier theorists on migration, such as E.G. Ravenstein, one of the early scholars to come up with a systematic theoretical study on migration, in his 1885 and 1889 papers, mentioned "unattractive climate" along with other factors as a reason for migration.[10] By 1994, a pattern was observed by Astri Surkhe who categorized research on environmental migration into two contradicting groups: maximalist and minimalist.[11] Minimalists emphasize that migration is not a monocausal phenomenon and that environmental change contributes to other general reasons for migration. Maximalists posit that environmental degradation is a direct cause of migration.[12]

In his 1985 book, Essam El-Hinnawi, an Egyptian Research Professor based in Cairo, and a maximalist, coined the term environmental refugees

[8] Sherbinin, Alex de, Kathryn Grace, Sonali McDermid, Kees van der Geest, Michael J. Puma and Andrew Bell "Migration Theory in Climate Mobility Research" *Systematic Review*, Frontiers, 10 May 2022 Vol. 4. https://ntrs.nasa.gov/api/citations/20220007965/downloads/MPumaFrontiersClimateMobilityReprint.pdf, p. 1. Accessed on 12 January 2023 Hunter, Lori M. and Daniel H. Simon (2022) "Time to Mainstream the Environment into Migration Theory?" *International Migration Review*. https://doi.org/10.1177/01979183221074343.

[9] Ibid.

[10] Ravenstein, E.G. (1889) "The Laws of Migration" *Journal of the Royal Statistical Society*, 52 (2), pp. 241–305, p. 286.

[11] Piscano, Marie Frances "Rethinking the Relationship of Migration, Environment and Peace and Security" *Peace and Progress-The United Nations Peace and Progress*. 3 (1), pp. 3–14.

[12] See, Ranjan, Amit "Climate Change, Environmental Migration and Population Displacement" in Amit Ranjan, Rajesh Kharat and Pallavi Deka (edited) *Environment, Climate Change and Migration in South Asia*. Routledge: New Delhi, pp. 12–30.

to refer individuals "who have been forced to leave their traditional habitat, temporarily or permanently, because of a marked environmental disruption (natural and/or triggered by people) that jeopardized their existence and/or seriously affected the quality of their life".[13] He categorized environmental refugees into three categories. First, those who are temporarily displaced because of environmental stress return after the situation is improved and the region is rehabilitated to its original State. Such a situation occurs with people suffer from natural hazards such as earthquakes, cyclones, or floods. The second category of environmental refugees is those who are permanently displaced to resettle in new areas because of permanent changes affecting their original habitat. This occurs in the case of building big dams and lakes, etc. The third category of environmental refugees is individuals or groups who migrate from their original home, temporarily or permanently, to a new place within the national boundaries, or in foreign countries, in search of a better quality of life. They migrate because the resource base in their original habitat has deteriorated to such a degree that it can no longer meet their basic needs.[14]

After El-Hinnawi, many other terms defining climate change-induced migrants have been coined or come into existence. First, environmental migrants are those who, "predominantly for reasons of sudden or progressive changes in the environment that adversely affect their lives or living conditions, are obliged to leave their habitual homes, or choose to do so, either temporarily or permanently, and who move within their country or abroad".[15] There is no international consensus on appropriate terminology, and the International Organisation for Migration (IOM) has provided working definition for environmental migrants.[16] Second, environmentally displaced persons are those "who are displaced within their

[13] El-Hinnawi, Essam (1985) *Environmental Refugees*. Nairobi: United Nations Environment Programme, p. 4.

[14] Ibid, pp. 4–5.

[15] International Organization for Migration (2014), "Migration, Environment, and Climate Change: Evidence for Policy Glossary", 13, https://publications.iom.int/system/files/pdf/meclep_glossary_en.pdf?language=en. Accessed on 19 June 2019.

[16] Md Shamsuddoha, S.M. Munjurul Hannan Khan, Sajid Raihan, Tanjir Hossain (2012), *Displacement and Migration from Climate Hot-spots in Bangladesh: Causes and Consequences*. ActionAid, Bangladesh, p. 18, https://unfccc.int/files/adaptation/groups_committees/loss_and_damage_executive_committee/application/pdf/displacement_and_migration_in_bangladesh.pdf.

country of habitual residence or who have crossed an international border and for whom environmental degradation, deterioration or destruction is a major cause of their displacement, although not necessarily the sole one".[17] The term "environmentally displaced person" is used as a less controversial alternative to environmental refugee or climate refugee.[18]

Whatever term(s) one uses, empirical studies show that people are forced to move out from their original habitat due to environmental reasons. Minimalists' and maximalists' approaches to studying climate change-induced migration depend on the region or country of study. Some countries or region within a country is more vulnerable to climate change compared to other. The next section examines climate-induced displacement in South Asia.

3 Climate-induced Migration and Displacement in South Asia

In South Asia, according to *ActionAid* and *Climate Action Network South Asia* report, by 2050, nearly 63 million people could be displaced from their homes.[19] *The report Groundswell: Preparing for Internal Climate Migration* (2018) projects that, by 2050, without concrete climate and development action, in South Asia, "internal climate migrants" could number over 40 million, representing up to 1.8 percent of the region's total population.[20] The share of climate migrants in all internal migrants could reach as high as 25 percent. Bangladesh's share may be above 50 percent because of its climate vulnerability. Climate out-migration hotspots occur in areas of deteriorating water availability and crop productivity; these areas include the northern part of the Gangetic Plain and

[17] International Organization for Migration (2014), "Migration, Environment, and Climate Change: Evidence for Policy Glossary", 13, https://publications.iom.int/system/files/pdf/meclep_glossary_en.pdf?language=en. Accessed on 19 June 2019.

[18] Ibid.

[19] Actionaid (2020, 18 December) "Climate Migration in South Asia Set to Treble by 2050 Due to Political Inaction on Global Warming." https://actionaid.org/news/2020/climate-migration-south-asia-set-treble-2050-due-political-inaction-global-warming. Accessed on 19 January 2022.

[20] World Bank Group (2018) "Internal Climate Migration in South Asia" Policy Note # 2. https://openknowledge.worldbank.org/bitstream/handle/10986/29461/GroundswellPN2.pdf?sequence=7&isAllowed=y. Accessed on 12 January 2022.

some areas of the broader Gangetic Plain and the corridor from Delhi to Lahore. In Bangladesh, movement will come from the rice-growing areas of the northeast. Out-migration hotspots also comprise low-lying lands, cities, and coastlines vulnerable to sea rise and storm surge impacts. However, urban areas will continue to support large and growing numbers of people. This dampening effect may be seen in major coastal metropolitan areas such as Mumbai, Dhaka, Chittagong, and Chennai. Rural-to-urban migration will generally remain strong. Several hotspots of climate in and out-migration are in transboundary areas. They include parts of Nepal and parts of western Bangladesh along the border with India.[21] According to the Global Report on Internal Displacement 2022, in 2021, nearly 5.3 million people were displaced due to natural disasters, lower than the region's decade average of 6.2 million. The decline is partly a result of a weaker monsoon season in the region.[22] India had 4.9 million displacement due to natural disasters.[23] Then, 99,000 in Bangladesh, 121,000 in Sri Lanka, and 25,000 in Afghanistan were displaced because of natural disasters in their respective countries.[24]

The IPCC study published in 2022 observes that if global warming is not limited to 1.5 degrees Celsius or at least 2 degrees Celsius, the consequences will be severe. The report identifies human activities as the cause of climate change and its implications, which include an increase in environmental migrants. Like other regions of the world, rising global temperature will increase heat waves, droughts, and cyclones in India. In 2022, India experienced its highest March and third-highest April temperatures in the last 122 years. The heatwaves affected 15 Indian states, including Himachal Pradesh, known for pleasant weather at this time of the year.[25] Due to human-induced climate change, by 2022, India's drought-prone area was increased by 57 percent since 1997.

[21] Ibid.

[22] Global Report on Internal Displacement 2022 *Internal Displacement Monitoring Centre*. https://www.internal-displacement.org/sites/default/files/IDMC_GRID_2022_HQ.pdf, pp. 67–68.

[23] Ibid.

[24] Ibid.

[25] Sharanya Hrishikesh & Meryl Sebastian "Delhi suffers at 49C as heatwave sweeps India" *BBC*, 16 May 2022. https://www.bbc.com/news/world-asia-india-61242341. Accessed on 12 July 2022.

Around 50 million people are affected by drought every year.[26] Simultaneously, an increase in temperature has led to a rise in sea level, which is projected to submerge some critical properties and road networks in India's coastal cities of Mumbai, Kochi, Mangalore, Chennai, Vishakhapatnam, and Thiruvananthapuram by 2050. There is also an increase in the frequency of cyclones in India. A study by scientists from the Indian Institute of Tropical Meteorology (IITM) has found a 52 percent increase in the frequency of cyclones over the Arabian Sea between 2001 and 2019. On the flip side, the study notes an eight percent decrease in the frequency of cyclones over the Bay of Bengal compared to the period between 1982 and 2002.[27] Sea-level rising will affect infrastructure in the coastal cities of India. For instance, in Mumbai, around 998 buildings and 24 kilometers of road length will be affected by potential sea-level rise by 2050, and approximately 2490 buildings and a road length of 126 km will be affected by a potential sea-level rise during high tide.[28]

To deal with the challenges of climate change, in 2009, the government of India came out with a National Action Plan on Climate Change. Subsequently, other measures were announced. In 2021, at the 26th Conference of Parties in Glasgow, Scotland, the Indian Prime Minister Narendra Modi declared a five-fold strategy called panchamrita (five ambrosia) to achieve net-zero emissions by 2070. The strategy includes[29]:

[26] "Desertification: 'Droughts Reduced India's GDP by up to 5% in 20 Years'" *Down to Earth*, 11 May 2022. https://www.downtoearth.org.in/news/climate-change/desertification-droughts-reduced-india-s-gdp-by-up-to-5-in-20-years--82794. Accessed on 19 July 2022.

[27] "52% Rise in Frequency of Cyclones in Arabian Sea Over Two Decades: Study" *Hindustan Times*, 18 July 2021. https://www.hindustantimes.com/india-news/52-rise-in-frequency-of-cyclones-in-arabian-sea-over-two-decades-study-101626550577184.html. Accessed on 19 July 2022.

[28] "Key Infra in Mumbai, Other Coastal Cities May Submerge by 2050" *Hindustan Times*, 8 April 2022. https://www.hindustantimes.com/india-news/key-infra-in-mumbai-other-coastal-cities-may-submerge-by-2050-101649357980809.html. Accessed on 12 May 2022.

[29] "CoP26: Modi Offers 'Panchamrita' Concoction for Climate Conundrum at Glasgow" *Down to Earth*, 2 November 2021. https://www.downtoearth.org.in/news/climate-change/cop26-modi-offers-panchamrita-concoction-for-climate-conundrum-at-glasgow-80001#:~:text=Modi's%20'Panchamrita'%20promises%20include%3A,one%20billion%20tonnes%20by%202030. Accessed on 11 May 2022.

1. India will get its non-fossil energy capacity to 500 gigawatts by 2030.
2. India will meet 50 percent of its energy requirements by 2030 with renewable energy.
3. India will reduce its projected carbon emission by one billion tonnes by 2030.
4. India will reduce the carbon intensity of its economy by 45 percent by 2030.
5. India will achieve net zero by 2070.

According to a report by the Government of Pakistan, the country is threatened by climate change in the following ways[30]:

1. Considerable increase in the frequency and intensity of extreme weather events, coupled with erratic monsoon rains causing frequent and intense floods and droughts;
2. Projected recession of the Hindu Kush-Karakoram-Himalayan (HKH) glaciers due to global warming and black carbon so to deposits from indigenous and transboundary pollution sources, threatening water inflows into the Indus River System (IRS);
3. Increased siltation of major dams caused by more frequent and intense rains in the catchment areas and subsequent floods;
4. Rising temperatures resulting in enhanced heat and water-stressed conditions, particularly in arid and semi-arid regions, leading to reduced agricultural productivity;
5. Increasing air pollution from agriculture, transport and industry resulting into smog inflicting huge loss to aviation, reduced mobility, loss of lives in accidents due to poor visibility and health hazard;
6. Further decrease in the already scanty forest cover from too rapid change in climatic conditions to allow natural migration of adversely affected plant species and wildlife habitat;
7. Increased intrusion of saline water in the Indus delta, adversely affecting coastal agriculture, mangroves and the breeding grounds of fish;

[30] Ministry of Climate Change Government of Pakistan, National Climate Change Policy, October 2021. https://mocc.gov.pk/SiteImage/Policy/NCCP%20Report.pdf. Accessed on 18 January 2023, p. 3.

8. Threat to coastal areas due to projected sea-level rise and increased cyclonic activity due to higher sea surface temperatures;
9. Increased stress between the upper and lower riparian regions in relation to sharing of water resources;
10. Increased health risks and climate change-induced migration.

In 2021, Pakistan had 70,000 people displaced due to natural disasters.[31] In 2022, severe floods directly or indirectly affected around 33 million people in Pakistan. 1730 people lost their lives, and, till October 2022, eight million displaced people faced health-related crises. The flood also pushed between 8.4 and 9.1 million more people below the poverty line.[32] Floods in Pakistan were mainly because of the impact of climate change. Global warming is rapidly melting glaciers in Gilgit-Baltistan and Khyber Pakhtunkhwa, creating more than about 3000 lakes. Out of them, 33 are at risk of sudden bursting, which could cause floods putting around seven million people at risk.[33] In 2022, due to climate change factors, Pakistan received around 390.7 mm of rain, nearly 190 percent more than its 30-year average from June to August.[34] To deal with the climate change's impact, Pakistan's National Climate Change Policy aims[35]:

1. To pursue sustained economic growth by appropriately addressing the challenges of climate change;
2. To integrate climate change policy with other inter-related national policies;

[31] Global Report on Internal Displacement 2022 *Internal Displacement Monitoring Centre*. https://www.internal-displacement.org/sites/default/files/IDMC_GRID_2022_HQ.pdf, pp. 67–68.

[32] The World Bank "Pakistan: Flood Damages and Economic Losses Over USD 30 billion and Reconstruction Needs Over USD 16 billion—New Assessment", 28 October 2022. https://www.worldbank.org/en/news/press-release/2022/10/28/pakistan-flood-damages-and-economic-losses-over-usd-30-billion-and-reconstruction-needs-over-usd-16-billion-new-assessme. Accessed on 12 January 2023.

[33] Georgina Rannard (2022, 2 September) "How Pakistan Floods are Linked to Climate Change" *BBC*. https://www.bbc.com/news/science-environment-62758811. Accessed on 12 January 2023.

[34] Ibid.

[35] Ministry of Climate Change Government of Pakistan, National Climate Change Policy, October 2021. https://mocc.gov.pk/SiteImage/Policy/NCCP%20Report.pdf, Accessed on 18 January 2023, p. 2.

3. To focus on pro-poor gender sensitive adaptation while also promoting mitigation to the extent possible in a cost-effective manner;
4. To build climate-resilient infrastructure;
5. To track impact of climate change on water, food, and energy security of the country, and to implement remedial plans to support water, energy, and food policies;
6. To minimize the risks arising from the potential increase in frequency and intensity of extreme weather events such as floods, droughts, and tropical storms;
7. To develop climate-resilient agriculture and food systems for all agro-ecological zones in the country;
8. To promote country's transition to cleaner, lower emission and less carbon intensive development;
9. To accelerate the policy coherence and integration to achieve the United Nations' Sustainable Development Goals (SDGs) in the light of its Sustainable Development Report 2020 (SDR2020) and our Nationally Determined Contributions;
10. To strengthen inter-ministerial and inter- provincial decision making and coordination mechanisms on climate change;
11. To facilitate effective use of the opportunities, particularly financial, available both nationally and internationally;
12. To foster the development of appropriate economic incentives to encourage public and private sector investment in adaptation and mitigation measures;
13. To enhance the awareness, skill, and institutional capacity of relevant stakeholders;
14. To promote tree plantation, conservation of natural resources, nature-based solutions and long-term sustainability.

Bangladesh faces a major climate change-related threat due to rising sea levels. Bangladesh is a member of the Climate Vulnerable Forum, a platform of the 48 most vulnerable countries in the world. Bangladesh is the world's seventh most climate-risk-affected country, though its contribution to global emissions is only 0.4 percent, calculated in 2018.[36] Climate

[36] Bangladesh: World Bank Group Country Climate And Development Report., 2022. https://openknowledge.worldbank.org/server/api/core/bitstreams/6d66e133-e49d-5ad9-b056-7b1a6c6206ed/content, p. 10. Also See National Adaptation Plan

change will lead to 13.3 million internal climate migrants in Bangladesh by 2050.[37] To deal with the climate change impacts, the Government of Bangladesh has framed some plans such as Bangladesh Delta Plan 2100, intending to achieve a safe, climate-resilient, and prosperous Delta, proposing US$38 billion by 2030 (in 2015 prices) for physical investments and institutional strengthening for implementation and monitoring.[38] In 2021, the Bangladesh government drafted the preliminary Mujib Climate Prosperity Plan, offering a multi-sectoral investment plan for climate-resilient infrastructure, clean energy, green value chains, and logistics from 2020 to 2030.[39] The preliminary plan estimates investments of US$89.7 billion until 2030. Bangladesh's Nationally Determined Contribution is approximately 2.2 percent of the gross domestic product (GDP) annually. The 2021 NDC Update identifies a requirement of US$32 billion to meet unconditional mitigation objectives and an additional US$143.8 billion to meet conditional objectives over 2021–2030. The total range of expenditure in all three plans is between 3.28 and 7.00 percent of the Bangladesh's GDP annually.[40]

The Maldives faces an existential threat due to the rise in sea levels. According to multiple reports from the National Aeronautics and Space Administration and the United States Geological Survey, almost 80 percent of the Maldives could become uninhabitable by 2050.[41] Sharing serious concerns and consequences of the climate change in the Maldives, at the Conference of Parties meeting in Glasgow, the president of the Maldives, Ibrahim Mohamed Solih, said, "...If we do not reverse this trend, the Maldives will cease to exist by the end of this century...If the rise in temperature remains unchecked at 1.5 and jumps to 2

of Bangladesh, 2022 (2023–2050), Ministry of Environment, Forest and Climate Change, Government of Bangladesh. https://www4.unfccc.int/sites/SubmissionsStaging/Documents/202211020942---National%20Adaptation%20Plan%20of%20Bangladesh%20 (2023–2050).pdf

[37] Ibid., p. 12.

[38] Ibid., p. 18.

[39] Ibid., pp. 18–19.

[40] Ibid., p. 19.

[41] Amit Ranjan "Rising Sea Levels: Threat for the Maldives" *ISAS's Briefs*, 8 December 2021. https://www.isas.nus.edu.sg/papers/rising-sea-levels-threat-for-the-maldives/.

degrees; that is a death sentence to the Maldives...".[42] 80 percent of the total islands of the Maldives are just a meter above sea level, making them vulnerable to sea waters. In 2008, the Maldives planned to divert a portion of its tourist revenue toward buying a new homeland. Then, president-designate Mohamed Nasheed said "We can do nothing to stop climate change on our own, and so we have to buy land elsewhere. It's an insurance policy for the worst possible outcome. After all, the Israelis [began by buying] land in [the] Palestine... We do not want to leave the Maldives, but we also do not want to be climate refugees living in tents for decades".[43] He also said that Sri Lanka and India were possible options as they have similar cultures, cuisine, and climate. Australia was also considered due to the availability of large tracts of unoccupied land.[44]

To deal with climate threats, the Maldives spends more than 50 percent of its national budget on adaptation. It is experimenting with 3-D printed coral structures or a floating city. The Maldives have worked on constructing artificial islands. An example is Hulhumalé. Construction of the island began in 1997, and the official settlement on the island was inaugurated in 2004. Reclamation projects have enlarged several other atolls. One of them is Thilafushi. Gulhifalhuea is the site of another land reclamation project. The Maldives is also planning to build sea walls.[45] To reduce emissions and dependence on fossil fuels, the Maldives aims to increase its renewable energy capacity to 85 MW by 2023. In 2020, the updated Nationally Determined Contribution of Maldives came out with a plan to reduce 26 percent of the country's emissions by 2030. Despite

[42] The President's Office, Republic of Maldives "Remarks by His Excellency Ibrahim Mohamed Solih, President of the Republic of Maldives at the 26th Session of the Conference of the Parties (COP26) to the United Nations Framework Convention on Climate Change (UNFCCC) held in Glasgow, Scotland" Remarks by His Excellency Ibrahim Mohamed Solih, President of the Republic of Maldives at the 26th Session of the Conference of the Parties (COP26) to the United Nations Framework Convention on Climate Change (UNFCCC) held in Glasgow, Scotland—The President's Office (presidency.gov.mv). Accessed on 19 January 2023.

[43] Randeep Ramesh (2008, 10 November) "Paradise Almost Lost: Maldives Seek to Buy a New Homeland" *The Guardian*. https://www.theguardian.com/environment/2008/nov/10/maldives-climate-change. Accessed on 18 June 2022.

[44] Ibid.

[45] Amit Ranjan (2021, 8 December) "Rising Sea Levels: Threat for the Maldives" *ISAS's Briefs*. https://www.isas.nus.edu.sg/papers/rising-sea-levels-threat-for-the-maldives/.

all such measures, the Maldives is highly dependent on other countries. The country's share in global GHG emissions is only 0.003 percent.[46]

According to the World Bank and the Asian Development Bank's Climate Risk Country Profile publication, by 2030, around 199,000 Nepalis may be affected by floods. At that time, the annual impact of floods on the country's GDP may be around US$574 million.[47] Simultaneously, there may be an increase in the number of droughts in Nepal. The National Climate Change Policy of Nepal, which replaced policy adopted in 2011, recognized that climate change affects the Nepali population's economy and livelihood. The goal of the policy is to take measures to build a climate-resilient society with following objectives[48]:

1. To enhance climate change adaptation capacity of persons, families, groups, and communities vulnerable to, and at risk of, climate change;
2. To build resilience of ecosystems that are at risk of adverse impacts of climate change;
3. To promote green economy by adopting the concept of low carbon emission development;
4. To mobilize national and international financial resources for climate change mitigation and adaptation in just manner;
5. To conduct research, make effective technology development and information service delivery related to climate change;
6. To mainstream or integrate climate change issues into policies, strategies, plans and programs at all levels of State and sectoral areas;
7. To mainstream gender equality and social inclusion (GESI) into climate change mitigation and adaptation programs

Despite such negative reports, growing effects, and policies, the Nepali government has hardly restrained itself from making decisions on projects

[46] Ibid.

[47] Referred in Amit Ranjan (2023, 3 January). "New Government in Nepal: Challenges and Concerns" https://www.isas.nus.edu.sg/papers/new-government-in-nepal-challenges-and-concerns/

[48] National Climate Change Policy, 2076 (2019) https://www.icimod.org/wp-content/uploads/2021/07/National-Climate-Change-Policy_english_2019_compressed.pdf. Accessed on 12 July 2022.

involving serious damage to forest areas. A recent example is the government's decision to build an airport at Nijgadh, 175 kilometers south of Kathmandu. According to an environmental and social impact assessment carried out by Nepal's tourism ministry in February 2017, about 2.4 million small and large trees had to be raged for the airport's construction. The matter went to the Supreme Court of Nepal, which quashed the decision taken by the government. The government formed an expert group in August 2022, pursuant to the Supreme Court's order to study the project. The expert group recommended abandoning the "airport city" concept to declare around 9000 hectares as a protected zone and to use only about 4000 hectares for the airport.[49] In 2022, the government demonstrated its concerns on environmental issues and signed a "concessional financing agreement" worth US$100 million with the World Bank to support "Green, Resilient, and Inclusive Development" in August 2022. This agreement has been opposed by many Nepali activists who argue that the country should receive climate finance instead of loans, given its minimum contribution to global emissions.[50]

Sri Lanka has experienced warming of around 0.8 °C over the twentieth century. Temperature rise has accelerated toward the end of the twentieth century. An average rise in temperature in Sri Lanka is projected to reach approximately 3.2 °C by the 2090s, less than the projected global rise of 3.7 °C. Maximum and minimum temperatures are projected to rise faster in Sri Lanka than the average but remain below global averages.[51] Sri Lanka is likely to experience more heat waves and an increase in heat

[49] Amit Ranjan (2023, 3 January). "New Government in Nepal: Challenges and Concerns" https://www.isas.nus.edu.sg/papers/new-government-in-nepal-challenges-and-concerns/

[50] Ibid.

[51] World Bank Group-Asian Development Bank "Climate Risk Country Profile-Sri Lanka", 2021. https://climateknowledgeportal.worldbank.org/sites/default/files/2021-05/15507-WB_Sri%20Lanka%20Country%20Profile-WEB.pdf, p. 11.

index and a number of days surpassing the heat index[52] of 35 °C.[53] Sri Lanka experiences many forms of flooding, such as river flooding, flash (or pluvial) flooding, and coastal flooding that affects several Sri Lankans. An increase in flooding will likely to increase the number of affected people between 70,000 and 560,000.[54]

In 2016, National Adaptation Plan for Climate Change Impacts in Sri Lanka was published. Its major goals are to[55]:

1. Raise the adaptive capacity of individuals, communities, and the society to cope with impacts of climate change effectively;
2. Reduce the vulnerability to climate risks by enhancing the resilience of communities and ecosystems, and;
3. Capture any opportunities that arise due to changes for maximum gain for the society and people.

The Objectives of the National Adaptation Plan for Climate Change are[56]:

1. To increase the resilience of economic sectors and natural systems against the emerging and projected impacts of climate change by adopting appropriate coping strategies and system improvements

[52] Heat index is also called apparent temperature. The heat index is what the temperature feels like to the human body when relative humidity is combined with the air temperature. There is direct relationship between the air temperature and relative humidity and the heat index. An increase or decrease in air temperature and relative humidity leads to increase or decrease in the heat index. See US Department of Commerce, National Oceanic and Atmospheric Administration, National Weather Service. https://www.weather.gov/ama/heatindex#:~:text=The%20heat%20index%2C%20also%20known,sweat%20to%20cool%20itself%20off. Accessed on 18 January 2023.

[53] World Bank Group-Asian Development Bank "Climate Risk Country Profile-Sri Lanka", 2021. https://climateknowledgeportal.worldbank.org/sites/default/files/2021-05/15507-WB_Sri%20Lanka%20Country%20Profile-WEB.pdf, p. 13.

[54] Ibid., p. 14.

[55] National Adaptation Plan for Climate Change Impacts in Sri Lanka, Climate Change Secretariat of Sri Lanka. https://www4.unfccc.int/sites/NAPC/Documents%20NAP/National%20Reports/National%20Adaptation%20Plan%20of%20Sri%20Lanka.pdf, Accessed on 19 January 2023, p. 47.

[56] Ibid., pp. 47–48.

2. To minimize the risk of damage caused by short-, medium-, and long-term impacts Climate Change Secretariat of Sri Lanka National Adaptation Plan for Climate Change in Sri Lanka 2016 48 associated with projected changes in climatic parameters through timely adaptive measures.
3. To expand the current knowledge on observed and projected changes of climate and associated physical vulnerabilities and socio-economic impacts through scientific research.
4. To build the capacity of communities, economic sectors, and ecosystems to adjust more readily to unfolding changes of climate through supportive investments on adaptive actions and increased awareness.
5. To improve the existing systems of disaster risk management to minimize the vulnerabilities and increase the risk preparedness for extreme events.
6. To increase the preparedness to face the threats of climate change through establishment of advanced monitoring and surveillance systems, timely weather and climate forecasting systems and effective communication channels for information dissemination.
7. To increase the skills and knowledge on successful practices of adaptation through well designed education, training, and awareness programs

Afghanistan faces warming rates higher than the global average, with a potential rise of 1.4 °C–5.4 °C by the 2080s and the 2090s.[57] Over the past century, precipitation in Afghanistan has varied. Some regions have witnessed a statistically high level, while others have seen the severity of drought, mainly in the southern part of the country, such as Kandahar, Helmand, and Nimruz.[58] According to INFORM (Index for Risk Management) 2019 index, Afghanistan is the fifth most natural disaster risk country in the world.[59] With the Taliban in power since August 2021, Afghanistan faces problems with competent people who can draw a plan and are technically qualified to implement measures to

[57] World Bank Group-Asian Development Bank "Climate Risk Country Profile-Afghanistan", 2021. https://climateknowledgeportal.worldbank.org/sites/default/files/2021-05/15396A-WB_Afghanistan%20Country%20Profile-WEB.pdf. Accessed on 21 January 2023.

[58] Ibid.

[59] Ibid.

deal with climate change-related issues. Afghanistan has been diplomatically isolated since 2021. Kabul was not invited to participate in the COP 27 held in 2022 at Sharm el-Sheikh. The only representative from Afghanistan was climate activist Abdulhadi Achakzai.[60]

4 Conclusion

As discussed in this chapter, climate change will affect many people in South Asia and force many to migrate internally; some may cross the border because of territorial contiguity. Not only people, climate change, as discussed in the chapter, affects the economy and infrastructure of the respective South Asian countries. All South Asian countries have developed policies to address imminent climate threats. However, their full implementation remains a problem. Second, the infrastructure works, such as in Nepal, mentioned above, create a dilemma where state has to opt between "development" and the environment. In many such cases, states opt for the former, creating a situation that aids emission.

Furthermore, as a contiguous territory, natural disaster in one country affects another. For example, floods in Bangladesh affect parts of the Indian State of West Bengal because of shared rivers.[61] In some cases, tensions are built over shared rivers. For example, Pakistan often accuses India of dry taps in the country.[62] Cross-border impacts of climate change need cooperation between the member states at bilateral and regional levels. In South Asia, tensions between India and Pakistan make regional cooperation difficult. South Asia's regional organization, South Asia Association for Regional Cooperation, is in suspended animation mode since 2016. Hence, any cooperation among South Asian countries to meet climate change-related challenges remains a problem.

[60] Ruchi Kumar (2022, 22 November) "Critics Lament Exclusion of Afghanistan from COP27" *Al Jazeera*. https://www.aljazeera.com/news/2022/11/22/experts-lament-exclusion-of-climate-risk-afghanistan-from-cop27. Accessed on 12 January 2023.

[61] "Bangladesh, India Floods Kill Over 100; Millions in Need of Aid" *Al Jazeera*, 2022, 21 June. https://www.aljazeera.com/news/2022/6/21/bangladesh-india-floods-kill-over-100-millions-in-need-of-aid.

[62] Safina Nabi and Syeda Sana Batool (2022, 17 October) "India's Dams and Pakistan's Water Crisis" *The Diplomat*. https://thediplomat.com/2022/10/indias-dams-and-pakistans-water-crisis/. Accessed on 12 February 2023.

CHAPTER 9

Priorities for Addressing Slow-Onset Events (SOEs) in Selected ASEAN Countries

Sivapuram Venkata Rama Krishna Prabhakar, Joy Jacqueline Pereira, Juan M. Pulhin, Nguyen Van Thang, Chhinh Nyda, Dannica Rose G. Aquino, Pham Thi Thanh Nga, Nguyen Dang Mau, Dang Quang Thinh, and Tran Thanh Thuy

Abstract Slow-onset events such as increasing temperature, desertification, loss of biodiversity and sea-level rise, and forest degradation constitute a significant part of climate change impacts that demand greater attention. Among these, ASEAN member states (AMS) are seriously concerned about sea-level rise and loss of biodiversity that can have

S. V. R. K. Prabhakar (✉)
Institute for Global Environmental Strategies (IGES), Hayama, Japan
e-mail: prabhakar@iges.or.jp

J. J. Pereira
Universiti Kebangsaan Malaysia's Southeast Asia Disaster Prevention Research Initiative (SEADPRI-UKM), Selangor, Malaysia
e-mail: joy@ukm.edu.my

V. Wijenayake et al. (eds.), *Linking Climate Change Adaptation, Disaster Risk Reduction, and Loss & Damage*,
https://doi.org/10.1007/978-981-99-8055-0_9

important implications for societal resilience and economic development, while other slow-onset events have localised impacts and significance. This chapter evaluates the national priorities in terms of specific slow-onset events, maps out various adaptation actions prioritised by countries, and assesses challenges that need to be overcome to realise these actions and provide a way forward. For this, the chapter utilises the experiences from Cambodia, Malaysia, the Philippines, and Vietnam.

Keywords Sea level rise · Biodiversity loss · Deforestation · Adaptation planning · Integrated solutions · Regional cooperation

Highlights

- The ASEAN countries are highly vulnerable to slow-onset events (SOEs). Despite the high vulnerability, the policy-level and ground-level focus on some of the SOEs is still lacking.
- SOEs are seen as beyond the capacity of individual countries to respond and mitigate. Countries see a great role in regional collaboration in understanding and addressing SOEs.
- The methodologies for isolating the impacts of SOEs need further progress. This needs to be strengthened from the point of view of data, monitoring and evaluation and linking the information to policy-level decision making. Synergies among SOEs demand the identification and implementation of integrated solutions.

J. M. Pulhin · D. R. G. Aquino
University of the Philippines Los Baños, Los Baños, Philippines
e-mail: jmpulhin@up.edu.ph

N. Van Thang · P. T. T. Nga · N. D. Mau · D. Q. Thinh · T. T. Thuy
Institute of Meteorology, Hydrology and Climate Change, Hà Nội, Vietnam

C. Nyda
Royal University of Phnom Penh, Phnom Penh, Cambodia

1 Introduction

ASEAN countries are highly vulnerable to a range of climate change-driven disasters and extreme events. Most of all, the emphasis is on typhoons that cause widespread devastation, and physical loss of human life, which become important media attention as well. However, the slow-onset events (SOEs) in ASEAN are equally important. Events such as sea level rise, deforestation, and loss of biodiversity are happening at a rapid pace in the region both due to climate change as well as human pressures. An increase in hot days is a major concern for the urban areas and an increase in the night temperature is a concern for the agricultural productivity in the region. The challenge of pinpointing some of these slow changes is partially masking the policy focus on these issues within the climate change domain. This signifies the need to reformulate the issue of SOEs for a proper policy focus in the region. Keeping this context in view, this chapter reviews the current evidence for SOEs in the ASEAN region and discusses them using a drivers, pressures, state, impacts, and responses (DPSIR) framework. While doing so, it utilises the results from the APN-funded project 'Integrating CCA, DRR and L + D to Address Emerging Challenges due to Slow-Onset Processes'. The legacy of the APN-funded project 'Integrating CCA, DRR and L + D to Address Emerging Challenges due to Slow-Onset Processes' has been that it proposed a risk framing approach that integrated disaster risk reduction, climate change adaptation, and loss and damages into a unified framework. Such an integrated risk framework has since been receiving growing attention at the global level and is very much relevant for the ASEAN region as a whole. Building upon this work, this chapter looks at the issue of SOEs from the lens of DPSIR framework that could communicate the message to stakeholders beyond climate change who may not be well verse with the risk and vulnerability frameworks.

2 Methodology

This paper reviews the current evidence for SOEs in the ASEAN region, presents the impacts of SOEs, and discusses various initiatives to mitigate SOEs. This objective of the paper is achieved through a literature review that was carried out using the literature databases including google scholar, Springer, Elsevier, etc. The keywords used for searchers include 'slow-onset', 'ASEAN', names of specific countries, 'sea-level rise', climate change, biodiversity, ecosystems services, and salinity. For presenting the research results, the paper uses the Drivers, Pressures, Status, Impacts, and Responses (DPSIR) framework that is well received by the members of the environmental faculty (Fig. 1).

The DPSIR framework proposes that environmental changes can be seen as a progression of implications of multiple facets of the society and economy and that all the environmental changes have discernible factors that drive the changes. These drivers could be considered as underlying factors, and these could be so broad in nature that most sectoral approaches may tend to ignore them considering them as outside their domain of expertise or domain of influence and hence missing the important opportunity of addressing underlying factors while targeting whatever the issues the interventions are addressing. Such an understanding is crucial for the SOEs since SOEs by nature tend to be influenced (either aggravated or mitigated) by various socio-economic and sociocultural factors and human activities that can exert pressure on the environment that could lead to overall exacerbation of impacts of the SOEs.

The vulnerability and capacity framework can also be considered a suitable framework for presenting the issues related to SOEs. While vulnerability and capacity framework has largely been well explored by the disaster risk reduction (DRR) and climate change adaptation (CCA)

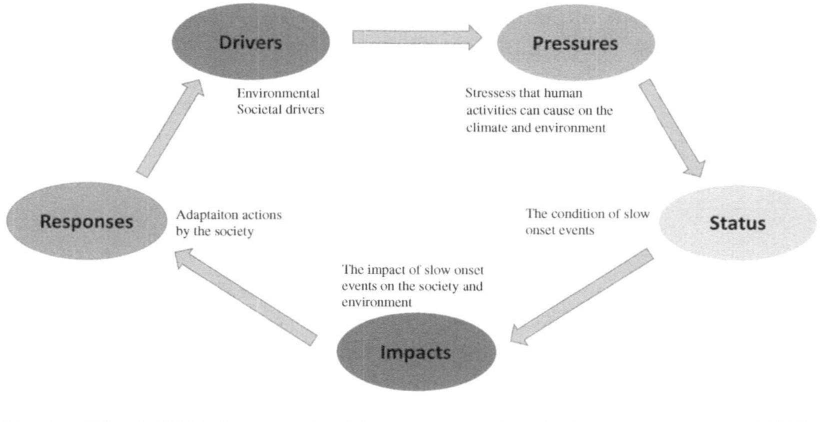

Fig. 1 The DPSIR framework of factors operating in slow-onset events (SOEs)

community, its use within the environmental community is negligible. In addition, vulnerability and capacity framework tends to work effectively at a much smaller scale while SOEs can pose multiple impacts on human society at a larger scale. Though this dimension is not well explored within the literature, it can be reasonably deduced that SOEs tend to take a larger shape in terms of the geographical area often spanning sub-regional and regional. Hence, thinking of SOEs from the DPSIR lens could provide an opportunity to explore the issue of SOEs from a broader perspective than the framework provides compared to the vulnerability and capacity framework. Such a perspective may also help communicate the issue to the wider environmental community that is stewards of environmental sub-sectors such as the urban environment, coastal zones, and forests that have direct stakes with the SOEs.

3 Results and Discussion

The thinking on slow-onset events (SOEs) within the climate change community can be said to have evolved alongside the discussions on loss and damage. It was primarily the Bali Action Plan (2007) of the Conference of Parties (COP) 13 at Bali, Indonesia (2007), that called for the need to focus on the loss and damage associated with climate change impacts (UNFCCC 2008). Subsequently, the debate on loss and damage has further gained momentum among climate change negotiators leading to the establishment of the Warsaw International Mechanism for Loss and Damage (WIM) at COP19, Warsaw, Poland (2013), to address the loss and damages associated with climate change including the slow-onset climatic events and extreme events especially focusing on developing and vulnerable countries. Accordingly, the slow-onset events (SOEs) have been referred to as the "*risks and impacts associated with increasing temperature, desertification, loss of biodiversity, land and forest degradation, glacial retreat and related impacts, ocean acidification, sea-level rise, and salinisation* (UNFCCC 2012)."

Subsequently, the Expert Group on Slow-Onset Events has provided a greater focus and meaning on how to approach the issue of SOEs. In its rolling plan of action, the Expert Group proposed that the issue of SOEs can be considered in a cascading fashion starting with the water-related SOEs such as desertification, glacial retreat, and sea level rise (UNFCCC 2021a, b). The idea has been that these three water-related SOEs can form the basis for the evolution of other SOEs such as salinisation, loss of biodiversity, land degradation and human mobility which could be considered cascading effects. Such a structured thinking of SOEs makes sense at a larger scale; however, many of the SOEs that are considered as cascading impacts of higher level SOEs could still happen independently of the higher level SOEs. This aspect needs to be explored further in the case of Southeast Asia. In the context of Southeast Asia, glacial retreat doesn't appear at the top of the priorities under SOEs. Hence, the focus of this paper would be on desertification and sea level rise as primary SOEs. Among the cascading SOEs, the paper will focus on salinisation, loss of biodiversity, land degradation and human mobility.

The work of the Expert Group has primarily been focused on strengthening the understanding of the SOEs at the regional and national scale (UNFCCC 2021a, b). Here, the two keywords are 'understanding' and the 'scale' that is focused on the regional and national levels. There are multiple reasons for this. First, the effects of SOEs are mainly considered to be manifested at a larger scale compared to sudden onset and extreme events such as typhoons and heavy rainfall events that focus on a much smaller scale.

Secondly, the scale of the SOEs is such that it demands regional cooperation be strengthened to address the capacity constraints of vulnerable countries. The focus on regional cooperation is especially important in regions such as Southeast Asia where the majority of the countries are highly vulnerable to climate change and are at the beginning stages of recognising and mitigating the impacts of SOEs. SOEs are often long-drawn processes and they require constant upgradation of capacities committed over a long period which could be considered by national governments as a burden, while they consider sudden onset events to demand immediate attention. Hence, in this chapter, an effort has been made to take a stock of the SOEs systematically using the DPSIR framework as discussed in the methodology section.

3.1 Drivers of Slow-Onset Events in the ASEAN Region

Climate change is the primary driver of SOEs. Climate change acts as a threat multiplier by exacerbating the social and environmental vulnerabilities that are unrecognised earlier. Climate change can primarily manifest in terms of sea level rise and desertification. Though sea level rise can happen due to land subsidence due to excessive extraction of groundwater, the combined effect of sea level rise due to land subsidence and climate change needs special attention in the ASEAN region especially due to the increasing urbanisation and related groundwater extraction. The greater development of urban areas along the sea coast also forms a major vulnerability for the region.

Climate change as a primary driver of SOEs could vary from country to country in the region. In Malaysia, a historical sea level rise of up to 7 mm per year has been recorded in some areas. Climate projections are available for the time slices of 2030 and 2050 covering temperature, rainfall, stream flow, evapotranspiration and soil moisture but these are unreliable for high-investment decisions (Pereira and Zain 2022). Using a new coastal elevation model CoastalDEM, it has been ascertained that nearly 31 million people in Vietnam, 23 million people in Indonesia, and 12 million people in Thailand will be living below the elevation of annual average flood in 2050 under the scenario of moderate emission cuts (RCP 4.5) and without factoring in the adaptation measures such as seawalls or levees (Climate Central 2019). Compared to the previous estimates that use SRTM data, the Climate Central work project an additional 51 million additional people under water by 2050. According to climate change scenarios in Vietnam, the sea level could rise by 100 cm by the end of the twenty-first century (MONRE 2022). By the end of the century, if the sea level rises by 100 cm due to climate change, the percentage of inundation risk are will be (i) 13.20% of the Red River Delta; (ii) 1.53% of the central coastal provinces from Thanh Hoa to Binh Thuan; (iii) 17.15% of Ho Chi Minh City and about 4.84% of Ba Ria—Vung Tau province; and (iv) 47.29% of the Mekong Delta (MONRE 2022).

Climate change is also affecting the health of ecosystems and their delivery of services. The analysis of local climate trends in Visayas, one of the major islands of the Philippines, revealed that the region underwent rising sea surface temperature consistent with the global trends, whereas the sea level projections for the country have been expected to exceed the global average (Mangaoang et al. 2019, 21–27). Warmer seas hold more carbon dioxide, which leads to acidification, and severe stress from the heat was the primary cause of the bleaching event that occurred in 1998 (Arceo et al. 2001, 587–590). In Luzon, elevated sea levels have already resulted in seawater intrusion in the Pampanga River, which was a precursor of salinisation and subsequent decline in soil fertility in rice farms (Lopez et al. 2004, 1).

The secondary drivers act as compounding agents and are equally important as that climate change. The secondary drivers can include increased coastal zone activities, increased water use for agriculture and other economic activities, unsustainable cropping and soil management choices that lead to soil salinisation. The secondary drivers here are discussed as those that are driven primarily by human activities. Many of these secondary drivers seem to collude with the pressures created by the primary drivers and hence are discussed in the section on the pressure below.

3.2 Pressures

We interpret pressures here as the stresses that SOEs create and their interaction with human activities. For example, mangroves play an important role in protection against sea level rise and related impacts. However, mangroves are also affected by other climatic events such as typhoons and human activities.

In the Philippines, the protection services from the mangrove ecosystem recently gained much attention after Typhoon Haiyan hit the Visayan island provinces of Samar and Leyte (Madid 2017, 2–3). However, widespread clearing of mangrove forests has been a problem since the 1960s, which was undertaken to accommodate agriculture and aquaculture expansion. As a result, a significant decline in mangrove cover became evident from 1967 until 1988, reaching an estimated loss of 8,000 ha per annum (FAO 2007, 23–26). Other environmental threats to mangroves and other coastal ecosystems include the proliferation of coastal residences and tourism activities that generate waste pollution. These problems were aggravated by the absence of programmes linking waste management to climate change adaptation and disaster risk management (Yoshioka et al. 2021, 13). Coastal damages caused by human activities have repercussions on the reproduction of aquatic species and the income of communities dependent on fisheries (Capili et al. 2005, 3).

In Malaysia, human activities in coastal areas are masking the effects of climate change on mangrove forest degradation and coastal erosion. Mangrove forests provide vital services including coastal protection, fish nurseries, water filtration, carbon sequestration, and eco-tourism opportunities. The mangroves are currently threatened by loss of habitat due to development activities in coastal zones. Sea level rise is expected to worsen this situation. A study conducted in the Sandakan area, Sabah, found that sea level rise has the potential to inundate about half of the current mangroves (MESTECC 2018). Coastal developments are beach resorts, ports and jetties, and important assets for economic growth. A total of 1349 km of coastline is continuously being eroded in the country. There are 44 areas with a total length of around 55 km delineated under the category of Critical Erosion and 309 areas with a total length of around 376 km classified in the Significant Erosion category. About 2344 areas with a total length of around 916 km are registered under the Acceptable Erosion category. Sea level rise poses additional risk to exposed structures, particularly along unprotected shorelines in the Critical Erosion and Significant Erosion categories.

While soil salinisation can happen due to the sea level rise and related saline water intrusion, the effects of soil salinisation can also be exaggerated due to human activities. Soil salinisation in the Philippines is being attributed to sea level rise, excessive pumping of groundwater, and infiltration of seawater into the river (Bureau of Soils and Water Management [BSWM] 2020, 3). Factors such as land use and vegetation suggest information on activities that may contribute to salinisation. It is estimated that 0.2 M ha of saline-prone coastal areas in the Philippines are significantly salt-affected, and the possible primary causes are linked to climatic factors (Philippine Rice Research Institute [PhilRice] 2001, 2). Specifically, these are associated with seawater intrusion caused by tidal fluctuations, inundation, and storm surges. Dry periods also induce salinity because less water is available to leach the salt from the soil surface leaving soils with high salt concentrations (Eswar et al. 2020, 310). Meanwhile, the secondary causes of salinisation are attributed to the pumping of saline irrigation water and groundwater and fertiliser residues.

Coastal areas are affected by saline intrusion in Vietnam. In recent years, water shortages and droughts have become more frequent in the coastal areas, especially in the Vietnamese Mekong Delta, leading to increasingly serious saline intrusion. In particular, drought and saltwater intrusion in the dry season 2015/2016 and 2019/2020 in the Vietnamese Mekong Delta, with saline intrusion inland up to 90–100 km in the dry season 2019/2020 (MOT 2021; VNMHA 2021). In addition, the increase in population, manufacturing activities, and industrial zones in coastal areas makes the demand for freshwater increase. Meanwhile, drought, runoff shortages, and saline intrusion are increasing. This is also a big challenge for coastal cities in Vietnam.

Managing scarce water resources is at the root of the fight against SOEs such as desertification. There are competing uses for limited water resources, and the competition is ever-growing in the ASEAN region and elsewhere. Agriculture, households, urban areas, industrial production, and energy production are some of the competing users of water resources. While the power sector has the dominant share of water abstracted, it uses water in a non-consumptive way (Philippine Statistics Authority [PSA] 2020, 2). The agriculture sector is the largest consumer of water in the Philippines, where a significant portion of water consumption is attributed to irrigation (World Bank, 2017). The growing population is a significant driver of increasing water demand. In 2003, around 63% of groundwater was extracted for household consumption, while 17% was utilised for agriculture (World Bank 2004, 27). However, in the same year, 60% of groundwater extraction was done without a permit (World Bank 2004, 27). This resulted in over-extraction and lowering of the aquifer triggering saline intrusion and ground subsidence (Greenpeace 2007, 27). Another driver of increased water consumption is fragmented management owing to the absence of a concrete institutional framework that is imperative to address issues and manage water resources. This fragmented approach caused the inefficient distribution of roles that inhibited sustainable water management (Barba 2005, 2).

Biodiversity is also facing conflicting interactions between humans and climate change in many parts of the ASEAN region. Terrestrial biodiversity is concentrated in forested land, which covered about 55% of Malaysia in 2014. The potential impacts of climate change on the biodiversity of terrestrial fauna are not widely investigated. The priority is on terrestrial fauna such as birds, orangutans, and elephants that are vulnerable to human-induced loss of habitat (MESTECC 2018). The contribution of climate change to land degradation is also sparsely recorded. Marine ecosystems are relatively better studied. The record indicates that rises in seawater temperatures have resulted in the bleaching of about 40% of the coral population and adversely affected the sex ratio of marine turtles. Climate change is projected to result in the degradation of mangroves and seagrasses, impacting fisheries.

3.3 Status of SOEs

In Malaysia, the combined impacts of sea level rise, storm surges, abnormally high tides, and high rainfall linked to severe flooding are poorly understood (MESTECC 2018). Climate hazards associated with floods and landslides are expected to become increasingly important as global warming proceeds to 1.5 °C (NADMA 2021). As the sea level rises, the potential inundation area of an earthquake-induced tsunami will expand further inland, increasing the exposure of physical assets and vulnerable communities. Thus, earthquake-induced tsunamis have an indirect link to sea level rise. The cascading effects of a fast-onset event such as floods could contribute to slow-onset environmental pollution, particularly in the coastal areas (Pereira et al. 2019; Yahaya et al. 2021). This could be further exacerbated by inundation and other coastal hazards that emerge due to sea level rise. An example of the worst-case scenario is depicted in a case study of western Selangor, where a landfill is expected to be submerged after the intensification of rainfall, coastal inundation under the extreme high tide, or a combination of both phenomena, accompanied by failed coastal protection. Cascading effects are expected on the health and well-being of surrounding communities, including damage to physical assets, degradation of mangrove habitats, and long-term productivity loss for the agricultural sector. While in this particular case, the point source for the environmental pollution is a landfill, in other areas it could be represented by an industrial facility housing toxic waste or a sewage treatment plant. Such scenarios could be widespread along the coast.

The Philippines is an archipelagic country and makes up a 36,289 km coastline (Licuanan et al. 2019, 515), the fifth longest in the world. Thus, elevated sea temperature and sea levels critically impact coastal and even inland ecosystems. Sea surface temperature significant rise was noted after the 1970s. Before this observation, the sea level already underwent a 0.15 m increase in 1940 (Cinco et al. 2013). The twentieth-century estimates of global sea level rise garnered a total of 19 cm for the period 1901–2010 (Church et al. 2013, 1191–1192). Areas found to be vulnerable to sea level were based on satellite observations and sensitivity measures including Leyte, Samar, and southern portions of Zamboanga. The highest sea level rise per year was observed in these areas between 1993 and 2015, which ranged from 4.5 to 5.5 mm (Kahana et al. 2016, 7–10). Future climate projections depict a temperature increase ranging from 1.8 °C to 2.2 °C. This will reduce precipitation and will make the months of March to May drier. On other hand, the southwest monsoon season is forecasted to cause extreme rainfall in Luzon and Visayas but this can be expected to decrease in Mindanao, while the predicted sea level will reach approximately 0.48–0.65 metres by 2100 (USAID 2017, 2).

Vietnam is experiencing various SOEs because of its long coastline of 3260 km and complicated topography. Most regions in Vietnam have been experiencing an increase in temperature, while sea-level rise threatens significant physical changes to the coastal zones of Vietnam. Along Vietnamese coastal areas, an upward trend in the mean sea level has been observed with an average increase of 2–3 mm per year (JICA 2015; MONRE 2021), while satellite altimetry data for the period 1993–2018 show an increasing sea-level trend of about 3.6 mm/year (Woillez and Espagne 2022).

3.4 Impacts of Slow-Onset Events in the ASEAN Region

The ASEAN region has a significant proportion of the population that is dependent on agriculture and coastal livelihoods. Though the proportion of the population dependent on agriculture and coastal livelihoods will likely decrease in the future as more and more people move away from natural-resource-based livelihoods, the importance of these activities in the overall economy and well-being of societies will continue to be there in the future. This necessitates that the impact of SOEs on agriculture and coastal livelihoods needs to be understood and addressed appropriately.

Further, increasing urbanisation and increasing linkages between urban and rural areas will likely expand the impact areas beyond rural areas and vice versa. This means that urban areas will bear the brunt of cascading impacts faced by rural food production centres. Isolating and buffering these shocks forms an important aspect of the strategy to address the impacts of SOEs.

Salinisation is on the rise, especially in coastal areas. The 2014 drought in Vietnam highlighted the compounding effects of drought on coastal salinisation and its implications on food security. The salinisation experienced in Vietnam showed a clear nexus between the severe drought that the country faced during the year. This indicated need to have coordinated and synergistic measures that address both the salinisation and drought impacts.

Climate variabilities in Cambodia are felt by the communities now. Many studies reported on people's perception of climate change and its implication on their livelihood (NCSD/MoE 2020). Many studies also indicated that climate change is going to shift the seasonal patterns mainly more rainfall in the wet season (May–Nov) and less in the dry season (Dec–April) (RGoC, Cambodia Climate Change Strategic Plan 2014–2023 2013a; Thoeun 2015). At the same time, the dry spell in the middle of the wet season in Cambodia in August, unique to Cambodia's climate context, will be longer (Nariddh 2021). With this dry spell and coupled with traditional rice cultivation, a long-rice variety that consumes the whole wet season period, will be greatly impacted, especially for subsistence farmers (Chhinh and Millington, 2015). It should be noted that many farmers in Cambodia are still relying on rain-fed cultivation and they are the most vulnerable to climate change impacts.

In Malaysia, sea level rise is projected to impact a major granary area in the low-lying coastal plains, northwest of the Peninsular, and affect coastal infrastructures such as ports and jetties. A total of 12 power plants, 30 transmission towers, 44 substations, and up to 7% of health facilities in the coastal areas of Peninsular Malaysia are reportedly exposed to sea level rise (MESTECC 2018). While the government reports that solid waste facilities are not projected to be impacted by sea level rise, research indicates that operating landfills and abandoned dump sites in coastal areas could give rise to problems (Yahaya et al. 2016, 2021). Salinisation of shallow groundwater aquifers is expected to contribute to water stress in some coastal areas and island communities (Hoque et al. 2016; Shamsudduha 2018; NAHRIM 2021); but the causes have not been attributed to human activities or sea level rise. The intrusion of saline waters is also expected to expand the distribution of brackish waters, which in conjunction with a warming trend would increase the risk and efficiency of malaria transmission.

With a long coastline of about 8840 km, Malaysia is experiencing climate hazards such as coastal erosion, coastal flooding and storm surges. For example, part of a jetty in Langkawi has been inundated during high tide since 2016 and the intrusion of saline water upriver has impacted a water treatment plant near the coast in Johor (MESTECC 2018). There has also been a breach of bunds that resulted in seawater intrusion, impacting some 63 ha of the fertile Muda Agricultural Development Authority (MADA) Granary, located in the low-lying coastal plains of northeast Peninsular Malaysia. The area is exposed to further seawater intrusion, especially during the southwest monsoon. While the attribution of the observed impacts to climate change is limited, sea-level rise, increase in sea surface temperature, and severity of storms will further exacerbate impacts on coastal communities and exposed infrastructure. There are areas including island communities that may be exposed to water stress.

Elevated sea levels and surface temperature result in the shift of reproduction and migration of species in the Philippines. Fisherfolk in villages located in Eastern Samar and Eastern Visayas experienced lower fish yield due to the declining quality of coastal and marine habitats (Mangaoang et al. 2019, 26–28). Fish vendors in Baybay, Leyte, noted that they are earning less due to the decline in fish catch. In addition to these impacts, seawater intrusion damaged crops and resulted in losses.

In Vietnam, low-lying areas including the Mekong river delta face major challenges from saline intrusion, which has already forced land-use changes, abandonment, and reduced yields in many provinces (World Bank, 2022; World Bank and ADB 2021a, b). In the Mekong river delta, during severe dry seasons, such as in 2016 and 2020 saline water has intruded up to 50 km inland from the coast and estuary, resulting in major crop damage (World Bank and ADB 2021a, b). According to a study by (MONRE 2021), the salt water of 4‰ in the low Mekong river delta is likely to intrude further 8.4–9. 5 km inland by 2050 under the RCP8.5.

The Mekong Delta is the largest rice production area in Vietnam, making an important contribution to socio-economic development and ensuring food security. However, the Mekong Delta is the region most affected and challenged by climate change. Among them, the biggest challenges include sea level rise causing loss of agricultural land, drought and salinity intrusion causing degradation of agricultural land, high temperature and changes in rainfall. The delta will face significant inundation risk due to sea level rise. According to climate change scenarios in Vietnam (MONRE 2022), sea level rise over the Mekong delta region is projected from 52 to 107 cm for RCP8.5 by 2100. Assuming 100 cm of sea level rise, 47.3% of the delta would face flood risks, including nearly 80% of Ca Mau Province and 75.7% of Kien Giang Province (Table 1). Its population and economy are most vulnerable to pronounced flooding threats related to sea level rise than any other region in Vietnam.

3.5 Responses to Slow-Onset Events

Any responses to SOEs in the ASEAN countries need to address the basic underlying vulnerability factors of these countries to SOEs. The Climate Vulnerability Forum (CVF) of the Global Center on Adaptation (GCA) has developed the third Climate Vulnerability Monitor (CVM3) that provides insights into more than 30 vulnerability indicators representing the environment, society and economy. Figure 2, derived

Table 1 Impacts of SOEs on the society, environment, and economy of ASEAN countries

Country	*SOEs*	*Major impacts*			*References*
		On society	*On environment*	*On economy*	
Cambodia	Sea level rise, salinity, forest loss, biodiversity loss etc	• Agricultural production • Houses damaged in coastal areas • Infrastructure damage • Migration	• Land degradation • Soil erosion • River bank erosion • Pest outbreak	• Income loss • Animal loss	MAFF (2021), MoE (2021)
Malaysia	Temperature rise Sea level rise Coastal erosion Saline water intrusion	• Shortage of drinking water on islands • Decreased rice production • Emergence of malaria	• Loss of biodiversity • Degradation of mangroves • Erosion of submerged dumpsites	• Loss of agriculture productivity • Loss of assets to coastal floods and storm surges	MESTECC (2018), Hoque et al. (2016), Shamsudduha (2018), NAHRIM (2021)
Philippines	Increasing sea surface temperature		• Extreme typhoons • Migration of species	• Decrease in fish yield • Limited food supply drives • An increase in prices of fish and shell	Comiso et al. (2015), Mangaoang et al. (2019)

(continued)

Table 1 (continued)

Country	*SOEs*	*Major impacts*			*References*
		On society	*On environment*	*On economy*	
	Sea level rise	• Coastal communities become more vulnerable to storm surges • Relocation of coastal residents	• Habitat destruction due to seawater intrusion	• Damages to livelihood and properties	Mangaoang et al. (2019)
	Ocean acidification	• Health-related impacts from toxic microorganism blooms	• Coral bleaching • Disruption of migration patterns	• Decrease in sources of marine-based livelihood	Mangaoang et al. (2019), Capili et al. (2005)
Vietnam	Increasing temperature, sea level rise, saline water intrusions	• Shortage of drinking water, forced immigration, public health, livelihoods, unemployment, work capacity	• Loss of biodiversity • Shrinking of mangrove forest	• Crop failure • Economic damage • Reduced productivity	MONRE (2021), Woillez and Espagne (2022)

from the freely accessible online CVM3 tool, shows the relative vulnerability of ASEAN countries to climate change in general on the Global Data Lab Vulnerability Index (GDL Vulnerability Index) (Huisman and Smits 2022). It shows various socioeconomic vulnerability factors such as economic growth, poverty, education, health, gender inequality, governance, demography, and access to basic infrastructure contribute to the climate change vulnerability in the ASEAN countries. On a 0–100 scale of the GDL Vulnerability Index, most ASEAN countries fall within the range of 15.3 (Singapore) to 54.8 (Lao PDR). Only Cambodia and Lao PDR cross the 50-point mark of the Index, while Thailand, Malaysia, Brunei, and Singapore, have vulnerability below 40 points. This indicates a wide range of socio-economic vulnerability of ASEAN countries to climate change. This signifies that probably solutions that work in one country may not work effectively in another country or that they need to be tailored to the socio-economic context of these countries.

Many of the socio-economic aspects covered by the above vulnerability index deserve responses that span from short-term to long-term

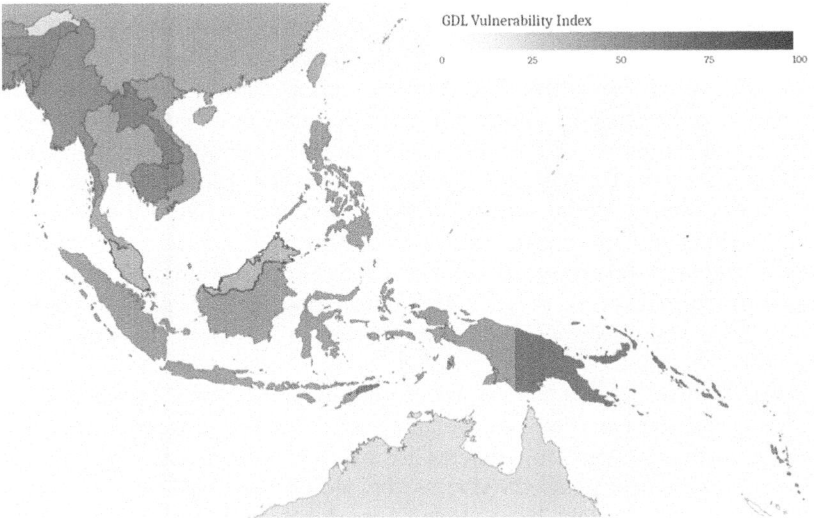

Fig. 2 The vulnerability index of the third Climate Vulnerability Monitor for the ASEAN region (*Source* Huisman and Smits 2022)

periods. This is because factors such as poverty, education, and health have a strong linkage with the developmental progression of these countries and hence may have to be dealt with within a short period. Issues such as demography and access to infrastructure could be addressed in the medium to long term. These long-term areas are also the ones that show a strong lock-in effect, i.e. any changes made in response to climate change would continue to stay for a considerably long period and frequent changes may not be feasible. Hence, countries need reliable climate change projections and an understanding of the adaptation benefits of actions that they implement in the short and long term. While the climate change projections are improving over the period, thanks to the efforts well-coordinated by the UNFCC and given impetus by the assessments carried out by the IPCC, the same cannot be said for the adaptation actions.

There is a huge gap in the way the climate change and impact projections are converted into identifying suitable adaptation solutions. The science of identifying and assessing adaptation solutions to mitigate location-specific climate change impacts is still in the nascent stages. The process is also complicated by not so well clarified and communicated concepts of resilience, disaster risk reduction, vulnerability, and risk making it even more challenging for the local stakeholders to effectively assess the available solutions. Questions such as should the stakeholders be assessing actions and solutions for the adaptation or vulnerability efficacy or risk reduction efficacy or aim at resilience are not clear. The same challenge can also be seen in the policy area as well. Climate change policies, as opposed to climate change actions that tend to have more physical interventions and social interventions, tend to have a wider area of influence and depending on what solutions they propose they may also have long-term implications as well. Adaptation policies also play an important role in scaling up adaptation by providing sufficient incentives and impetus to them.

Despite their importance in governing adaptation at a broader scale, the progress in adaptation policymaking is still in the nascent stages. For most countries adaptation, policies are mainly enshrined in the form of national adaptation plans (NAPs) as the global focus on national adaptation plans has largely been driven top down with the push coming from the UNFCCC Conference of Parties (COP), the Paris Agreement, Cancun Adaptation Framework, and subsequent negotiations. To date, among all ASEAN countries, only Cambodia submitted the NAP, while the rest of the countries are in the process of developing. However, the formulation of sub-national adaptation plans, such as city adaptation plans, and sectoral and provincial adaptation plans, may be at a much more advanced stage than the national adaptation plans. It could take several more years before most ASEAN countries complete their NAPs and for them to incentivise sub-national adaptation plans. ASEAN also being one of the active inter-governmental bodies at the regional level, also needs to develop a regional adaptation plan and efforts are well underway with discussions to shape such a plan already begun. For example, the ASEAN has already created an 'ASEAN Regional Plan of Action for Adaptation to Drought 2021–2025' (ASEAN 2021a, b).

Completing regional and national adaptation plans will help countries to guide adaptation actions in a much more coherent manner. More importantly, they will have good guidance on focusing on issues that need long-term and sustained emphasis as opposed to focusing on short-term and immediate outcomes that may be short-lived. Short-term responses often provide a false sense of security and dependency and hence there is a need to identify those measures that can have long-term impacts.

From the point of view of moving from a short-term focus to a long-term, issues such as monitoring and evaluation (M&E) of adaptation play an important role as well. For this, M&E of mitigation actions could provide a good starting point as ASEAN countries have already made significant progress in M&E of mitigation actions and this learning can guide the adaptation as well. Some of the challenges involved in M&E of adaptation actions are also related to the challenges of measuring adaptation or adaptation metrics (Prabhakar and Srinivasan 2011). Though several ideas have been proposed to measure adaptation, the major limiting factor for operationalising these metrics appears to be a lack of clear contexts such as adaptation benchmarks or adaptation targets or adaptation goals against which the progress in adaptation can be measured and reported. While developing such metrics and corresponding adaptation goals could take time, countries such as Indonesia are at the forefront of developing climate change vulnerability indices that they propose to use as a means of tracking adaptation over a period. These bottom-up experiences emerging in the region could be considered for wider application in the region.

Regional cooperation appears to be an essential aspect of the fight against SOEs in the ASEAN region. As the region is already ripe for such regional cooperation with initiatives such as the ASEAN Working

Group on Climate Change (AWGCC), and the proposed work plan of the AWGCC can usher the region into strong regional cooperation on climate change. The 2014 drought has in particular made the region realise such as strengthening cooperation in responding to climatic events. There are already proposals to prepare an action plan for climate change adaptation and mitigation for the ASEAN region (ASEAN 2021a, b). Realising such proposals would go a long way for the region to strengthen cooperation to mitigate SOEs as well.

One of the least realised needs of the region is in the area of natural resource monitoring. With more emphasis on nature-based solutions, the need to monitor the biophysical resources in the ASEAN region receives even more attention. Further, the looming threat of loss of biodiversity, deforestation and calls to incorporate natural resources accounting into the Gross Domestic Productivity (GDP) strengthens the need to even more closely monitor biophysical resources in the region. While deforestation monitoring is well-developed, the monitoring of biodiversity loss and ecosystem function loss is not well-developed.

Countries such as Myanmar that heavily rely on natural resources (e.g. mining) for their economic prosperity are at the forefront of the benefactors of such monitoring. This is also an important area where the country could cooperate at the regional level. Closely related issues are the issue of macro- and micro-level data that can support various aspects of climate change risk management including strengthening climate change projections, impact assessments and climate change forecasts. The improved weather data are also necessary to strengthen the medium- and long-term weather predictions that can help farmers to take appropriate choices well in time. While measures to strengthen severe weather forecasting are rapidly underway, the same level of emphasis is required in the area of SOEs as well (WMO 2021).

Integrated coastal zone (ICZ) management has evolved as an important measure for addressing a wide range of issues such as coastal degradation, mangrove restoration, tsunami risk reduction, and protecting coastal erosion. Though the meaning and implementation mechanisms vary from country to country, ICZ provides a strong example of how integrated solutions can be made successful when employed in a limited context as in the case of coastal areas. Such integrated solutions are also required to address issues such as land degradation beyond coastal zones, soil salinisation and sea level rise. Regional bodies such as the Mekong River Commission (MRC) and Partnerships in Environmental Management for the Seas of East Asia (PEMSEA) on biodiversity are well placed to contribute to developing integrated solutions in the ASEAN region.

In Malaysia, efforts are underway to strengthen climate change adaptation with a focus on the management of water resources and security, coastal resources, agriculture and food supply, urban and infrastructure resilience, public health, forestry and biodiversity, and cross-sector approaches (GOM 2021). Systematic observations are conducted to monitor hydrologic parameters, drought information, and sea level changes. Models are also being developed to integrate sea level rise, storm surges, abnormally high tides and high rainfall for early warning and assessment of future scenarios (MESTECC 2018). The inputs will be channelled into Integrated Shoreline Management Plan and coastal development plans. Minimum platform levels are being established for new developments in the coastal areas to increase the resilience of coastal facilities. High-resolution coastal inundation maps and comprehensive assessments for major sectors are planned to identify clearer options for adaptation, including nature-based solutions and green-grey infrastructure (GOM 2021). Improved water management such as the building of check dams and tube wells, and early warning systems in coastal peat lands have contributed to reducing the occurrence of forest fires during dry spells and El Niño events. Outreach activities are focused on the conservation of the coastal environment, peat land and peat swamp forests, including replanting of mangroves, as well as food and water security, among others. Such informal environmental education is conducted in collaboration between public agencies, non-governmental organisations, and the private sector. More recently, there is a concerted effort to transform the water sector, which offers an opportunity to strengthen the integrated river basin management and integrated flood management approaches, for more effective land use planning in the country (EPU 2021; GOM 2021).

Biodiversity management is also being strengthened through the enhancement of the national forest inventory and monitoring of the marine environment as well as conservation efforts. In Peninsular Malaysia, the conservation of the Central Forest Spine has contributed to enhancing the resilience of forests to climate change. In Borneo, a major transboundary conservation programme initiated in 2007 by the Governments of Brunei, Indonesia, and Malaysia, in collaboration with non-government organisations, has contributed to improving community welfare as well as reducing deforestation, forest degradation, and the associated loss of biodiversity and ecosystem services. Forest enrichment is a major focus, where tree planting activities are conducted in logged forests and degraded areas using native species. In addition, sustainable agriculture development is also a vehicle for enhancing the environment and biodiversity in the country. Marine biodiversity resilience is being enhanced through the control of overfishing, prevention of habitat loss, habitat rehabilitation, biodiversity restoration, the establishment of additional protected areas, and enforcement of regulations to maintain critical habitats such as mangroves, seagrasses, and coral reefs. Active conservation of marine turtles is also conducted through the establishment of hatcheries in the country.

The adaptation measures that can help address SOEs in the Philippines should focus on the proper delineation of coastal areas to suitable uses, and programmes that will generate alternative livelihood to support stakeholders who are dependent on marine resources for livelihood and subsistence.

Rehabilitation of degraded coastal areas such as mangroves and beach forests is a relevant strategy to protect coastlines against SOEs and extreme climate conditions. To help vulnerable communities adapt to the impacts of SOEs, information, education, and communication (IEC) campaigns should be strengthened to build the awareness of the communities.

For decades, Vietnam has enacted various programmes, policies, and legislations to adapt to climate change and address its impacts. For example, laws related to climate change adaptation including Laws on Fisheries, Law on Cultivation, Law on Husbandry, Law on Biodiversity, Law on Environmental Protection, Law on Forestry, Law on Natural Disaster Prevention and Control, Law on Natural Resources and Environment of Sea and Islands, Law on Hydrometeorology, Law on Water Resources and Law on Dikes have been approved (Vietnam Gov 2020, 2022). In addition, the National Adaptation Plan (NAP) on climate change and the National Climate Change Strategy (NCCS) was approved by Decision 1055/QD-TTg dated 20/7/2020 and Decision 896/QD-TTg dated 26/7/2022, respectively. Currently, the National Action Plan Responding to Climate Change by 2030 is under processing for approval to concretise actions to implement the strategic tasks defined in the NCCS.

Given the mandate, line ministries and provincial-level governments have developed specific climate change policies at their respective levels such as climate change action plans, plans for implementation of the Paris Agreement and integrating climate change in sectoral development strategies (Vietnam Gov 2022). Typical adaptation models have also been developed and replicated, such as flood-resistant housing in many provinces in the Central region and a small-scale seawater desalination system in Ca Mau province, and salt-tolerant crops have been successfully bred in Mekong delta provinces (Vietnam Gov 2020). Moreover, many programmes and projects to adapt to climate change have been implemented showing uninterrupted efforts of Vietnam in climate change adaptation (Table 2).

4 Major Challenges

Despite the progress in adapting to SOEs presented in the previous section, the ASEAN region still faces some challenges. Institutions play an important role in the fight against SOEs we observe that there is a limited focus of institutions on some of the SOEs such as salinisation, ocean acidification, and sea level rise in the region. Issues such as ocean acidification are found to be beyond the jurisdiction and capacity of many countries to address by themselves and they see the need for coordinated action, especially at the global level.

Though some progress has been made in the monitoring land use changes, especially with a focus on forests, one of the major challenges in addressing the slow-onset events is the limited progress in comprehensive bio-physical monitoring systems in ASEAN countries covering water resources, forests, land, and biodiversity. The focus is especially lacking in the area of biodiversity monitoring and ecosystem services. Technology applications in this area are on the rise but still at nascent stages, such as remote sensing applications for biophysical resources, monitoring of drought and sea level rise, and early warning systems, etc. The limited data and quality of data mean limited use of available data for policy decision-making, a major challenge for many developing countries. Limited regional cooperation on the issue of data sharing and data standardisation in specific needs to be promoted for data harmonisation for designing regionally usable information and database and solutions.

Several methodological issues need focus. Slow on-set events are highly confounded with other biophysical and human processes, so isolating cause-effect relationships is challenging. Rigid sector-focused assessments often do not capture these changes. Developing case studies that isolate

Table 2 Major adaptation measures identified by national governments in the ASEAN region

Country	*SOE issue (SLR, desertification, migration etc.)*	*Short-term adaptation measures*	*Long-term adaptation measures*	*References*
Cambodia	Desertification	Building more irrigation schemes	Farmer agriculture knowledge and skills	MoE, Cambodia's Third National Communication (2020)
		Early warning system	Changing agricultural practices	
		Changing crop calendar/crop variety with consume less water	Investing in resilient crops	
		Safe migration		
Malaysia	Temperature rise	Systematic observations	Updated standards	MESTECC (2018), NAHRIM (2021), GOM (2021)
	Sea level rise	Modelling inundated areas	Minimum platform levels	
	Coastal erosion	Improved water management Early warning	Land use planning	
	Saline water intrusion	Conservation	Sector-specific options	
		Rehabilitation		
Philippines	Increasing sea surface temperature	Conduct post-disaster needs assessment	Implementation of national gendered ecosystem vulnerability and risk assessment	NDRRMC (2020)
		Provision of suitable relocation sites	Implementation of knowledge management on climate change and disaster	

Country	*SOE issue (SLR, desertification, migration etc.)*	*Short-term adaptation measures*	*Long-term adaptation measures*	*References*
		Construction of resilient housing Implementation of income-generating activities for alternative livelihood Improvement in the structural design of coastal infrastructure	Institutionalisation of natural resource accounting Development of financing plan for water sector climate change action plan	Vision 40 NEDA (2020)
Vietnam	Sea level rise, increasing temperature, droughts, saline intrusions		Promulgating National Adaptation Plan (NAP) Promulgating National Strategy on Climate Change (NSCC)	Decision 1055/QD-TTg dated 20/7/2020 Decision 896/QD-TTg dated 26/7/2022

human interventions with that of climatic changes in assessing the impacts and progression of SOEs is important.

There is no systematic monitoring of slow-onset events in many countries. There have been some introductions to the use of the Standardised Precipitation Index (SPI) to monitor drought but it is just in the pilot phase among non-governmental organisations (NGOs) for their interventions. The application of these developments to desertification is still unknown. Desertification, as a slow-onset disaster, is very hard to address from many different aspects. First, it is very hard to know the onset of the event. There are many efforts to develop early warning systems in Cambodia but yet to materialise (UNDP 2022). The ability to provide early warning for desertification and agricultural production would require long-term forecasting so that farmers can decide which crop variety they would produce during the season or change to another type of crop so that it is matching to the anticipated rainfall. Second, understanding desertification is still limited from academic to policy development; as a result, it is remaining in development challenge in the region (RGoC 2013b).

The current climate projections in Malaysia are unreliable and have to be applied judiciously, particularly concerning high investment decisions (Pereira and Zain 2022). Other approaches have to be developed to identify potential climate change impacts so that appropriate adaptation measures can be identified. Coastal hydrodynamic simulation needs to be enhanced to incorporate sea level rise impacts for vulnerable coastal regions in Malaysia (MESTECC 2018). Design standards and development guidelines in coastal areas must be enhanced to enable proactive adaptation to sea level rise. Systematic observations and research are insufficient on the potential impacts of climate change on forests and biodiversity; this aspect is not well incorporated as a component of Sustainable Forest Management in the country. In addition, the country has yet to develop a comprehensive National Adaptation Plan. An adaptation plan that integrates elements of disaster risk reduction from the Sendai Framework and the Sustainable Development Goals is critical to guide the systematic implementation of no-regret adaptation measures for all sectors. Lack of capacity and awareness of the full chain of implications of climate change impacts by key stakeholders in all major sectors, is also a major challenge.

Addressing SOEs in Vietnam is still facing some major shortcomings, particularly in the implementation of adaptation policies and in translating the guiding principles into practice. Firstly, efforts to respond to climate-related hazards often focus on short-term emergency actions and recovery, with limited attention paid to long-term prevention and proactive response activities. Secondly, hard measures are dominant compared to soft measures. Additionally, although most ministries and provinces have developed and/or updated their climate adaptation plan, the implementation of these plans has been very limited due to financial deficits. Another gap is the lack of stakeholder participation in adaptation planning and implementation, especially the limited engagement of local communities and/or commune-level government (Espagne et al. 2021). Also, there is a lack of regulations, incentive mechanisms and solutions to improve the role and position of women in policy formulation and implementation of climate change response (Vietnam Gov 2022).

5 Conclusion

This chapter reviewed the slow-onset climate events (SOEs) in the selected ASEAN countries inspired by the APN project entitled 'Integrating CCA, DRR and L + D to Address Emerging Challenges due to Slow-Onset Processes' implemented in Malaysia, the Philippines, Cambodia, Myanmar, and Vietnam. Using a DPSIR framework, the chapter laid out drivers and pressures, status, impacts of SOEs and responses of countries to SOEs. It attempted to provide an alternative framework to take a stock of the situation of SOEs in the region. It is evident from the discussion that the vulnerabilities of ASEAN countries to climate change fall on a spectrum with countries such as Lao PDR and Cambodia being highly vulnerable to climate change. While climate change is the major driver of SOEs in the region, the chapter has laid out the human factors that aggravate the SOEs and their impacts. SOEs such as sea level rise and salinity are highly interrelated and hence adaptation measures could seek synergies as well.

ASEAN has a strong regional mechanism in place and the mitigating SOEs could benefit from the regional cooperation. The cooperation should benefit countries in the areas of data sharing and data harmonisation. Strengthening natural resource monitoring can provide benefits beyond the adaptation such as in the area of natural resource accounting for GDP. The synergies in the area of M&E of adaptation are possible with the strengthening of databases. M&E of changes in biophysical resources could benefit strengthening nature-based solutions while addressing the loss of biodiversity and ecosystem services.

Countries also need to develop methodologies that isolate confounding effects and provide structured cause-effect relations in impacts. This can provide clarity in responding to SOEs as SOEs tend to last long-term with ample interaction with other human-induced pressures. Land use planning is receiving an impetus in the region with countries such as the Philippines integrating climate change projections into land use decisions and planning. Related measures in coastal zone regulations will effectively mitigate SOEs related to seal level rise, saline water intrusion, salinisation, etc. Land use planning can also promote policy and planning strategies that integrate CCA, DRR and loss and damage.

Slow-onset events are not as well understood as sudden events such as typhoons; hence they tend to receive less attention among academics, policymakers, and farmers. In Malaysia, the integration of climate change adaptation and disaster risk reduction is a major focus of the national development plan. Capacities and mechanisms are being strengthened to reduce disaster risks, loss, and damage and improved the management of socio-economic and environmental impacts. The transformation of the water sector, with emphasis on integrated river basin management, is an important vehicle to pursue land use planning to manage both fast and slow-onset climate hazards. The management of water resources and security, coastal resources, agriculture and food supply, urban and infrastructure resilience, public health, forestry, and biodiversity are also important aspects that can be considered using this approach. Climate change has been appearing rapidly at the national-level discourse in Vietnam, and this is expected to accelerate in the twenty-first century. Countries are increasingly seeing human activities such as urban development, population growth, and migration increasing the vulnerability to SOEs. To proactively mitigate and enhance resilience to the impacts of SOEs, it is recommended to develop mechanisms to enhance the participation of various stakeholders in the policy processes and implementation through a constant dialogue mechanism, and to mobilise financial resources to support climate change adaptation.

References

Arceo, H., M. Quibilan, P. Aliño, G. Lim, and W. Licuanan, 2001. *Coral bleaching in Philippine reefs: Coincident evidences with mesoscale thermal anomalies*. Bulletin of Marine Science, 69: 579–593.

ASEAN, 2021a. *ASEAN regional plan of action for adaptation to drought 2021–2025*. Jakarta, Indonesia: ASEAN.

ASEAN, 2021b. *ASEAN state of climate change: Current status and outlook of the ASEAN region toward the ASEAN climate vision 2050*. Jakarta, Indonesia: ASEAN, 190.

Barba, P., 2005. *Challenges in water resource management in the Philippines*. National Water Resources Board.

Bureau of Soils and Water Management, 2020. *National mapping, characterization and development of spatial database for the coastal areas affected by salinity*. The Philippines: Bureau of Soils and Water Management.

Capili, E.B., A.C.S. Ibay, R. Ramon Villarin, and J. Ramon, 2005. *Climate change impacts and adaptation on Philippine coasts*. Vol. 3, 2299–2306. https://doi.org/10.1109/OCEANS.2005.1640108.

Chhinh, N., and A. Millington, 2015. *Drought monitoring for rice production in Cambodia*. Climate, 3 (4): 792–811.

Church, J., P. Clark, A. Cazenave, J. Gregory, S. Jevrejeva, A. Levermann, M. Merrifield, G. Milne, R. Nerem, P. Nunn, A. Payne, W. Pfeffer, D. Stammer, and A. Unnikrishn, 2013. *Sea level change*. Cambridge, United Kingdom and New York, NY, USA: Cambridge University Press, Chapter 13, 1137–1216.

Cinco, T.A., F.D. Hilario, R.G. De Guzman, and E.D. Ares, 2013. *Climate trends and projections in the Philippines*. Paper presented at the 12th National Convention on Statistics (NCS), Mandaluyong City, Philippines.

Climate Central, 2019. *Flooded future: Global vulnerability to sea level rise worse than previously understood*. Princeton, United States: Climate Central, 12. https://juneau.org/wp-content/uploads/2019/11/2019-11-05-Bill-Leighty-ClimateCentral-FloodedFuture-2019CoastalDEMReport.pdf.

Comiso, J., G. Perez, and L. Stock, 2015. *Enhanced Pacific Ocean sea surface temperature and its relation to Typhoon Haiyan*. Journal of Environmental Science and Management, 18: 1–10. https://doi.org/10.47125/jesam/2015_1/01.

EPU, 2021. *Twelfth Malaysia plan 2021–2025*. Putrajaya: Economic Planning Unit, Prime Minister's Department.

Espagne E. (ed.), T. Ngo-Duc, M.-H. Nguyen, E. Pannier, M.-N. Woillez, A. Drogoul, T.P.L. Huynh, T.T. Le, T.T.H. Nguyen, T.T. Nguyen, T.A. Nguyen, F. Thomas, C.Q. Truong, Q.T. Vo, and C.T. Vu, 2021. *Climate change in Vietnam: Impacts and adaptation. A COP26 assessment report of the GEMMES Vietnam project*. France: AFD.

Eswar, D., R. Karuppusamy, and S. Chellamuthu, 2020. *Drivers of soil salinity and their correlation with climate change*. Current Opinion in Environmental Sustainability, 50: 310–318.

Food and Agriculture Organization, 2007. *The World's Mangroves 1980–2005. A thematic study prepared in the Framework of the Global Forest Resources Assessment*. Paris: FAO.

GOM, 2021. *Malaysia's update of its first nationally determined Contribution*. Putrajaya: Submitted to the UNFCCC, Ministry of Environment and Water.

Greenpeace, 2007. *The state of water resources in the Philippines*. Quezon City, Philippines: Greenpeace Southeast Asia.

Hoque, M.A., P.F.D. Scheelbeek, P. Vineis, A.E. Khan, K.M. Ahmed, and A.P. Butler, 2016. *Drinking water vulnerability to climate change and alternatives for adaptation in coastal South and South East Asia*. Climatic Change, 136 (2): 247–263.

Huisman, J., and J. Smits, 2022. *Projections of climate change vulnerability for the 21st century: The GDL vulnerability index (GVI)*. GDL Working Paper 22–101, Nijmegen, The Netherlands: Global Data Lab.

JICA, 2015. *Country report Vietnam: Natural disaster risk assessment and area business continuity plan formulation for industrial agglomerated areas in the ASEAN region*. Japan: JICA.

Kahana, R., R. Abdon, J. Daron, and C. Scannell, 2016. *Projections of mean sea level change for the Philippines*. United Kingdom: Met Office.

Licuanan, W.Y., R.W. Cabreira, and P.M. Aliño, 2019. *The Philippines*. In World seas: An environmental evaluation, 515–537. https://doi.org/10.1016/b978-0-08-100853-9.00051-8.

Lopez, M., V. Teodoro, and C. Mendoza, 2004. *Management options for salt-affected rice-based farming systems during the El Niño and La Niña phenomena*. Journal of Sustainable Agriculture, 23 (4): 19–37. https://doi.org/10.1300/J064v23n04_04.

Madid, C.M., 2017. *Increasing climate resiliency of Philippine mangroves through Impact Assessment*. IA's Contribution in Addressing Climate Change 37th Annual Conference of the International Association for Impact Assessment, 4–7 April 2017. Conference Proceedings. Montreal, Canada.

MAFF, 2021. *Annual report of agriculture, forestry, and fisheries 2021 and directions for 2022*. Phnom Penh: Ministry of Agriculture, Forestry and Fisheries.

Mangaoang, E.O., H.R. Montes, S.A. Cesar, and K.C. Fiecas. 2019. *Impacts of climate change SOEs of sea level rise, increasing sea surface temperature, and ocean acidification in coastal fishery areas and small island ecosystems in Visayas*. Philippines.

MESTECC, 2018. *Third national communication and second biennial update report to the UNFCCC*. Putrajaya: Ministry of Energy, Science, Technology, Environment and Climate Change.

Ministry of transportation (MT), 2021. *Update climate change factors into the plan for the development of transportation infrastructure in the Vietnamese Mekong Delta up to 2030, vision 2050*. Technical report of Ministry Project. Ha Noi, Vietnam.

MoE, 2020. *Cambodia's Third National Communication*. Phnom Penh: Ministry of Environment (MoE).

MoE, 2021. *The 4th state of environment report*. Phnom Penh: Ministry of Environment (MoE).

MONRE, 2021. *National Climate Assessment*. Vietnam Publishing House of Natural Resources, Environment and Cartography, Ministry of Natural Resources and Environment.

MONRE, 2022. *Climate change scenarios for Vietnam*. Vietnam Publishing House of Natural Resources, Environment and Cartography.

NADMA, 2021. *National risk register*. Putrajaya: National Disaster Management Agency, Prime Minister's Department of Malaysia.

NAHRIM, 2021. *Climate change adaptation framework for water sectors*. Seri Kembangan: National Water Research Institute of Malaysia, Ministry of Environment and Water.

Nariddh, M.C., 2021. *Digging deep to irrigate rice fields during dry spell*. Accessed 11 12, 2021. https://www.khmertimeskh.com/50806745/digging-deep-to-irrigate-rice-fields-during-dry-spell/.

National Disaster Risk Reduction and Management Council (NDRRMC), 2020. *National disaster risk reduction and management plan: 2020–2030*. Camp Aguinaldo, Quezon City: Office of Civil Defense–Policy Development and Planning Service (OCD–PDPS).

National Economic and Development Authority (NEDA), 2020. *Vision 2040: The Filipino public consultations: Discussions with the Filipino youth*. Manila, The Philippines: NEDA.

NCSD/MoE, 2020. *A third study on understanding public perceptions of climate change in Cambodia: Knowledge, attitudes, and practices*. Phnom Penh: Ministry of Environment (NCSD/MoE).

Pereira, J.J., and M.K. Zain, 2022. *Climate change adaptation in Malaysia*. In J.J. Pereira, M.K. Zain, R. Shaw (eds). Climate change adaptation in Southeast Asia. Singapore: Springer, 103–116.

Pereira, J., J. Pulhin, N. Chhinh, D. Tran, and S. Satari, 2019. *Appraising slow onset hazards for loss and damage: Case studies in Southeast Asia*. APN Science Bulletin, 9 (1). https://doi.org/10.30852/sb.2019.720

Philippine Rice Research Institute, 2001. *Management of salt-affected soils for rice production*. Department of Agriculture. 2001 No. 40.

Philippine Statistic Authority, 2020. *Country's total water abstraction increased by 12.3 percent from 2010 to 2019*. Press Release Reference No. 2020–295. Manila, The Philippines.

Prabhakar, S.V.R.K., and A. Srinivasan, 2011. *Metrics for mainstreaming adaptation in agriculture sector*. In Rattan Lal, Mannava V.K. Sivakumar, S.M.A. Faiz, A.H.M. Mustafizur Rahman and Khandakar R. Islam (eds). Climate change and food security in South Asia. Dordrecht: Springer, 551–567.

RGoC, 2013a. *Cambodia climate change strategic plan 2014–2023*. Phnom Penh: Ministry of Environment.

RGoC, 2013b. *National action plan for disaster risk reduction* (NAP-DRR) 2014–2018. Phnom Penh: NCDM.

Shamsudduha, M. 2018. *Impacts of human development and climate change on groundwater resources in Bangladesh*. Groundwater of South Asia, 523–544.

Thoeun H.C., 2015. *Observed and projected changes in temperature and rainfall in Cambodia*. Weather and Climate Extremes 7 (March): 61–71.

UNDP, 2022. *Strengthening climate information and early warning systems to support climate-resilient development in Cambodia*. Accessed 11 12, 2022. https://www.adaptation-undp.org/projects/ldcf-ews-cambodia.

UNFCCC, 2008. *Report of the Conference of the Parties on its thirteenth session, held in Bali from 3 to 15 December 2007*. Addendum. Part Two: Action taken by the Conference of the Parties at its thirteenth session. Decisions, Bonn, Germany: UNFCCC, 60. https://unfccc.int/documents/5079#beg.

UNFCCC, 2012. *Slow onset events*. Bonn, Germany: UNFCCC, 61. https://unfccc.int/resource/docs/2012/tp/07.pdf.

UNFCCC, 2021a. *Expert group on slow onset events: Rolling plan of action*. Bonn, Germany: UNFCCC, 2. https://unfccc.int/sites/default/files/resource/POA_SOEs.pdf.

UNFCCC, 2021b. *Five-year rolling workplan of the Executive Committee of the Warsaw International Mechanism for Loss and Damage associated with Climate Change Impacts*. Five year rolling workplan, Bonn, Germany: UNFCCC, 20. https://unfccc.int/sites/default/files/resource/5yr_rolling_workplan.pdf.

United States Agency for International Development, 2017. *Climate change risk in Philippines: Country fact sheet*. The Philippines: USAID.

Vietnam Gov, 2020. *Vietnam nationally determined contributions*. The socialist republic of Vietnam.

Vietnam Gov, 2022. *Vietnam nationally determined contributions* (updated in 2022). The Socialist Republic of Vietnam.

Vietnam Meteorological and Hydrological Administration (VNMHA), 2021. *The hydro-meteorological characteristics in 2020 in Vietnam*. Ha Noi, Vietnam.

WMO, 2021. *Strengthening severe weather forecasting in South and Southeast Asia*. Geneva, Switzerland: WMO. https://public.wmo.int/en/media/news/strengthening-severe-weather-forecasting-south-and-southeast-asia.

Woillez, M.-N., and Espagne, E. (ed.), 2022. *The mekong delta emergency, climate and environmental adaptation strategies to 2050*. Final Report GEMMES Vietnam project. Paris: Agence Française de Développement.

World Bank and ADB, 2021a. *Climate risk country profile*. Vietnam: The World Bank Group and the Asian Development Bank.

World Bank and ADB, 2021b. *Climate risk country profile*. Philippines. The World Bank Group and the Asian Development Bank.

World Bank, 2004. *Philippines environment monitor: Assessing progress* (English). Washington, DC: World Bank Group.

World Bank, 2017. *Philippines urbanization review: Fostering competitive, sustainable and inclusive cities*. Full report. Washington, DC: World Bank Group.

World Bank, 2022. *Vietnam country climate and development report*. CCDR Series. Washington, DC: World Bank. © World Bank. https://openknowledge.worldbank.org/handle/10986/37618. License: CC BY 3.0 IGO.

Yahaya, N.S, J.J. Pereira, and M.R. Taha, 2021. *Using best available information to conduct impact assessment of future climatic hazards on a landfill*. Climate Change, 165 (68): 1–19.

Yahaya, N.S., L. Choun-Sian, M.R. Taha, and J.J. Pereira, 2016. *Exposure of municipal solid waste disposal sites to climate related geohazards: Case study of Selangor*. Bulletin of the Geological Society of Malaysia, 62: 57–63.

Yoshioka, N., M. Era, and D. Sasaki, 2021. *Towards integration of climate disaster risk and waste management: A case study of urban and rural coastal communities in the Philippines*. Sustainability, 13 (4): 1624. https://doi.org/10.3390/su13041624.

CHAPTER 10

SWOC Analysis on the Proposed Linkage Between Climate Change Adaptation (CCA), Disaster Risk Reduction (DRR) and Loss and Damage (L&D): Case Studies in the Low-lying Coastal Cities of Indonesia, Philippines, Thailand and Vietnam

Hop T. B. Hoang, Rodel D. Lasco, Ninh H. Nguyen, and Rizaldi Boer

Abstract Climate change impacts, particularly climate-related disasters, are increasingly affecting Southeast Asia. This chapter addresses the linkages between Climate Change Adaptation (CCA), Disaster Risk Reduction (DRR) and Loss and Damage (L&D) in low-lying coastal cities of Indonesia, the Philippines, Thailand and Vietnam. The project proposed

H. T. B. Hoang (✉) · N. H. Nguyen
Center for Environment, Research, Education and Development (CERED), Hanoi, Vietnam
e-mail: h.hophp@gmail.com

V. Wijenayake et al. (eds.), *Linking Climate Change Adaptation, Disaster Risk Reduction, and Loss & Damage*,
https://doi.org/10.1007/978-981-99-8055-0_10

an integration framework to support risk governance in these countries and beyond. The chapter reviews L&D assessment activities in the research countries, critically analyses the proposed framework and suggests further improvements for better application of integrating DRR, CCA and L&D to address climate change challenges in the area. The chapter highlights the need for an integrated approach to address climate-related disasters, which can contribute to reducing L&D in the vulnerable communities of Southeast Asia. The chapter emphasises the importance of coordinated efforts and cooperation among different stakeholders, including governments, communities and experts, to achieve effective risk governance and enhance resilience in the face of climate change.

Keywords Loss and damage · Climate change adaptation · Disaster risk reduction · Climate-related disasters · Resilience

Highlights

- It is necessary to link DRR, CCA and L&D to cope with climate change issues and to be more resilient to climate-related risks in the countries, helping mainstream climate change aspects in to the policy and development activities
- Existing frameworks collect data on disaster-related loss and damage in these countries, but the lack comprehensive coverage of intangible costs inherent in the Loss and Damage (L&D) concept.

N. H. Nguyen
e-mail: nguyenhuuninh1954@gmail.com

R. D. Lasco
Oscar M. Lopez Center for Climate Change Adaptation and Disaster Risk Management Foundation, Inc., Pasig City, Philippines
e-mail: rlasco@omlopezcenter.org

R. Boer
Centre for Climate Risk and Opportunity Management, IPB University, Dramaga, Indonesia
e-mail: ccrom_rizaldi@apps.ipb.ac.id

- To implement the framework, there is a need to strengthen the database, stakeholder engagement, financial assistance, risk governance, communication methods, technical capacities and specific assessment.

1 Introduction

Southeast Asia (SEA), including the Philippines, Vietnam, Thailand, Indonesia and Cambodia is more vulnerable to the effects of extreme climate-related events, including flooding, sea level rise and tropical storms and drought. Climate change-induced alterations in rainfall, temperature, sea-level, and disaster patterns are anticipated to significantly exacerbate vulnerabilities in the region, especially within the agriculture sector and low-lying coastal cities (IPCC, 2022, chapter 6; Dodman et al., 2022). Additionally, given their positioning in low-lying coastal areas with densely populated regions, cities in this area are projected to experience substantial rises in average economic losses from flooding between 2005 and 2050 (Dodman et al., 2022). Extreme rainfall due to tropical depressions/storms cause serious floods in these countries and significantly damages infrastructure (urban and rural), livelihoods (agriculture), freshwater availability, health and the properties of local citizens (Harmeling & Eckstein, 2013). Additionally, slow-onset events such as extreme temperatures and the urban–heat island effect also increase the vulnerability of those cities to climate change (Dodman et al., 2022).

For decades, DRR and CCA have been seen as probable solutions to climate problems to address the negative impacts of climate change, including climate-related disasters. CCA is considered as an adjustment process to actual and/or projected climate and its effects, in order to lessen harm, including harm relating to climate-related disasters, and/or increase beneficial opportunities (IPCC, 2007). Meanwhile, DRR supports preventing new and reducing existing disaster risks and managing residual risks UNISDR (2017). DRR is a process aimed to lessen risks from climate–induced or non-climate-induced sudden–onset events (Kreienkamp & Vanhala, 2017). DRR and CCA overlap significantly. Hence, in the context of the increasing impact of climate-related issues, the linkage between DRR and CCA is believed to enhance the resilience of communities and attain sustainable development goals (Shaw et al., 2016).

However, it is also believed that under the gravity of the current climate situation, uncertainty of the future change of climate and complicated socio-economic systems, decision makers also need to address the definition of loss and damage. The term "loss and damage" expresses the manifestation of issues as a result of climate change in developing countries (UNFCCC, 2008), which cannot be avoided by efforts of the CCA (UNFCCC, 2014) and could be averted by mitigation (UNFCCC, 2013; 2016). They are unavoidable adverse effects of climate change occurring when the impacts of climate change go beyond the ability of people/systems to deal with and/or to adapt to (Huq et al., 2013)—the "residual impacts". The impacts are caused from both sudden and/or slow-onset events. As assessing the impacts of these events is not straightforward, people talk about economic and non-economic loss and damage. Meanwhile, Loss and Damage (L&D) is aimed at showcasing policies and strategies addressing loss and damage in the UNFCCC Warsaw International Mechanism. Then, it is understood that L&D is meant to express a comprehensive process that begins with climate change mitigation and extends to the application of climate change adaptation, including the integration component of CCA linkages with DRR to address problems of climate-related disasters, then culminates in demonstrating L&D resulting from unavoidable climate change issues.

L&D could help to unite CCA and DRR solutions to reduce the negative impacts of climate change. The information on loss and damage is important for decision-makers, policymakers and other stakeholders to comprehend and implement appropriate actions and legal documents. Loss and damage under assessments relating to DRR, such as post-disaster damage and needs assessments, would provide information on climate-related disasters CCA programmes; meanwhile, CCA programmes could receive and send information on slow-onset events of climate change for DRR activities to address climate change-related disaster. Since the UNFCCC Warsaw International Mechanism, L&D has received more attention from society and can be used in climate change negotiations, and international/national efforts can be established to promote the role of L&D in DRR and CCA efforts for L&D in DRR and CCA 's more climate-resilience strategies.

In order to support international mechanisms to facilitate regional coordination on CCA, DRR, and L&D, APN has supported a series of scientific research and capacity development efforts in linking CCA, DRR, and L&D by way of its Climate Adaptation Framework. The research projects on the Linkage between Climate Change Adaptation (CCA), Disaster Risk Reduction (DRR), and Loss and Damage (L&D):

Case Studies in the Low-lying Coastal Cities of Indonesia, Philippines, Thailand and Vietnam, is one of those projects. Based on the results of the research project, this paper will (1) provide an overview of L&D assessment activities with climate change and disaster risks in the research countries, (2) critically analyse the proposed framework to link DRR, CCA and L&D, and (3) propose additional suggestions for a better application of the integration linkages between DRR, CCA and L&D to effectively address climate change challenges in the area.

2 Methodology

2.1 *Literature Review*

A large number of legal documents relating to DRR and CCA as well as related reports and research in Indonesia, Philippines and Vietnam were reviewed in the from 2014 to 2022 to address two main questions: (1) Are there frameworks for assessing climate risks in these countries, including loss and damage? and (2) Review existing frameworks for assessing L&D due to climate-related disasters.

2.2 *Focus Group Discussions*

During the period 2014–2016, at least two national workshops were conducted in each country, facilitating focus group discussions with representives from government agencies, academia, local authorities and other stakeholders involved in an activity relating to DRR, CCA and L&D in Philippines, Indonesia and Vietnam. This method yields detailed information as well as stakeholders perspectives on the topic. Furthermore, it helps validate key findings from literature reviews and discuss the potential to link DRR, CCA and L&D proposing a framework to integrate these concepts.

2.3 *Key Informal Interview*

In November and December 2022, an open-ended questionnaire was prepared for key informal interviews with academia, representatives of several NGOs and government agencies in Vietnam, the Philippines and Indonesia as well as experts in DRR and CCA activities. Interviews were conducted. The interviews were aimed at analysing the Strengths (S), Weaknesses (W), Opportunities (O) and Constraints (C) (SWOC) of applying a proposed framework to link DRR, CCA and L&D.

3 Results and Discussion

3.1 Current Loss and Damage System in Researched Countries in Southeast Asia

3.1.1 Overview Research on Loss and Damage in Southeast Asia

Loss and Damage (L&D) is a new concept in terms of the residual risk of climate change. An existing framework for L&D assessments is not likely to be available at the regional or national level. However, there are relevant initiatives and/or mechanisms in regions/sectors/countries. These initiatives/mechanisms vary in scope, time, place and type of climate stressors and approaches (Lasco et al., 2017). National post-disaster assessment systems for DRR activities in researched countries are a significant example of the mentioned mechanisms/initiatives. Generally, the systems generated both monetary (damaged houses/cars/boats/roads/factories, etc.) and non-monetary (deaths, injuries, etc.) negative impacts of (climate-related) disasters.

Related post-disaster assessment systems for DRR activities could be viewed as an example for a L&D system. The systems are designed to collect both monetary and non-monetary negative effects of climate-related disasters, including extreme weather events, which may be directly computable or not. Data from post-disaster assessments serve to reduce disaster risks (DRR), aiding in identifying needs in impacted areas/regions to support their first-aid, rehabilitation, and recovery (UN ECLAC, 2014).

Additionally, the negative impacts of past extreme weather events also provide input for policymakers to develop legal documents, programs, and projects on CCA, as climate change is protected to increase risks from such events in the long term. For example, historical data on the impact of climate-related disasters and the uncertainty surrounding climate change serve as inputs when the governments of OECD countries develop projects, legal documents, and programs through a "climate lens". Southeast Asia countries such as the Philippines, Vietnam, Indonesia and Thailand are encouraged to consider historical data on the impact of climate-related disasters and possible trends of climate change scenarios when developing national, regional and sectoral programs and projects, and legal documents. Climate proofing is a compulsory requirement from donors such as the Asia Development Bank and World Bank when those countries need support from them.

Based on the literature review and key informal interviews conducted during the period 2014–2016 for the project Linkage between Climate Change Adaptation (CCA), Disaster Risk Reduction (DRR) and Loss and Damage (L&D): Case Studies in the Low-lying Coastal Cities of Indonesia, Philippines, Thailand and Vietnam, existing systems and tools related to post-disaster assessment systems/disaster database systems for L&D assessments in Vietnam, Indonesia and the Philippines are presented in Table 1.

Along with more than 180 other countries, the researched countries implement Sendai Framework for Disaster Risk Reduction, and it is evident that there is a need to assess the responding needs of affected places. Hence, the countries have existing systems relating to post-disaster assessment. Generally, the information is collected from the local to central level (for example, the system in Indonesia, Vietnam) that could be used for DRR planning in the affected areas. For the more specific assessments relating to both DRR and CCA activities in these countries, assessments would be conducted by assessment team(s), including representatives from sectoral agencies and stakeholders (for example, experts, academia, etc.). This information would not only be used for quick relief but also for further recovery and further planning activities.

At the moment, not only state actors (for example, military, central government and ministries) take part in disaster risk management activities in the countries but also non-state actors (for example, communities, private firms, civil society, non-government organizations, academia, etc.) (Lassa, 2013; Lasco et al., 2017). Non-state actors would not only participate in damage assessment activities, but also utilize information from government reports to promptly aid local communities and instigate further planning for DRR and CCA activities.

3.2 SWOC Analysis of the Proposed Regional Framework of Linkage Between DRR, CCA and L&D

3.2.1 Overview of Suggestion and the Proposed Regional Framework

According to Lasco et al. (2017), DRR and CCA activities should implement a framework that has a holitsic approach to integrating L&D information into the system addressing climate impacts/disasters through CCA and DRR strategies. The framework is shown in Fig. 1. The proposed framework emphasises the importance of L&D assessment systems in improving the effectiveness of DRR & CCA activities in consideration of current and potential challenges and opportunities. To specify:

Table 1 The L&D assessment system in collaborating countries

Countries	*Assessment tools and purposes*	*Components & Key actors*
Philippines	– Post-Disaster Needs Assessment (PDNA) to assess the (financial, technical and human) demands for "recovery, reconstruction and risk management".	– Includes 3 main parts: (1) Damage Assessment and Loss Assessment (DALA) caused by disasters. (2) Human Recovery Needs Assessment (HRNA), an (3) Proposal of Recovery Framework for immediate recovery within 2 years and long-run growth beyond 3 years. – Key actors: Local Government Units, sectoral agencies, the Office of Civil Defense (OCD). The National Disaster Risk Reduction and Management Council (NDRRMC), PDNA team, and cabinet cluster.
Vietnam	– Damage Assessment and Need Assessment (DANA) to assess physical and financial damage, and assess needs for quick relief and compensation to disaster-stricken communities.	– Includes 2 main parts: (1) Physical and financial damage assessment: in 13 sectors. (2) Needs assessment include immediate relief needs, post-disaster and recovery needs. – Data is collected from local to central levels via the Committee for Natural Disaster Prevention and Control (CNDPC) and General statistics Office (GSO)*—according to Law on Disaster risk management and Joint Circular No. 43/2015/ TTLT-BNNPTNT-BKHDT dated Nov 23, 2015, on guidance for producing statistics of, and assessing damages caused by, natural disasters.

(continued)

Table 1 (continued)

Countries	*Assessment tools and purposes*	*Components & Key actors*
	Disaster Information Management System (DesInventar) to report disaster impacts in Vietnam.	The data is collected from the local to the central level via the systems of the Committee for Natural Disaster Prevention and Control (CNDPC) and the General Statistics Office (GSO)*. Then the information on loss and damage inputs into the (DesInventar) system.
	Socioeconomic Impact Assessment (SEIA) to assess the socioeconomic impacts of intangible factors which would support figuring out the resilience and/or recovery ability of research places (APEC, 2009).	SEIA would help to fill the gap of other assessment frameworks (especially social gaps) in DRR and CCA activities.
Indonesia	– Risk Retention and Risk Transfer (pilot project) to provide emergency aid, social safety nets and contingency fund/loan which would help to transfer risk (eg. risk in agriculture—farming) through an insurance mechanism.	Key actors: Government agencies (Ministry of Agriculture, National Boards of Disaster Management, Military Forces), Non-Government Organizations, academia, private sector.
	Disaster database system (Data Informasi Bencana Indonesia, DIBI) has been developed basing on the DesInventar system since 2006. It includes communication forums that help to (1) collect, validate and store disaster data (UNDRR, 2020); (2) integrate risk mapping and risk-informed planning. Hence, it would support localizing DRR planning. It would provide information for risk identification, policy formulation, and funding prioritization based on disaster trends in the local areas. Post-Disaster Needs Assessment (PDNA) and similar assessments were also applied for tsunamis, earthquakes, and some flooding cases (Jibiki, 2020) for years.	Local Disaster Management Agencies (BPBD) (the data is collected from central—local); BRR (Badan Rehabilitasi dan Rekonstruksi:The Agency for the Rehabilitation and Reconstruction of Aceh/Nias) (Jibiki, 2020). Non-government Organizations, academia.

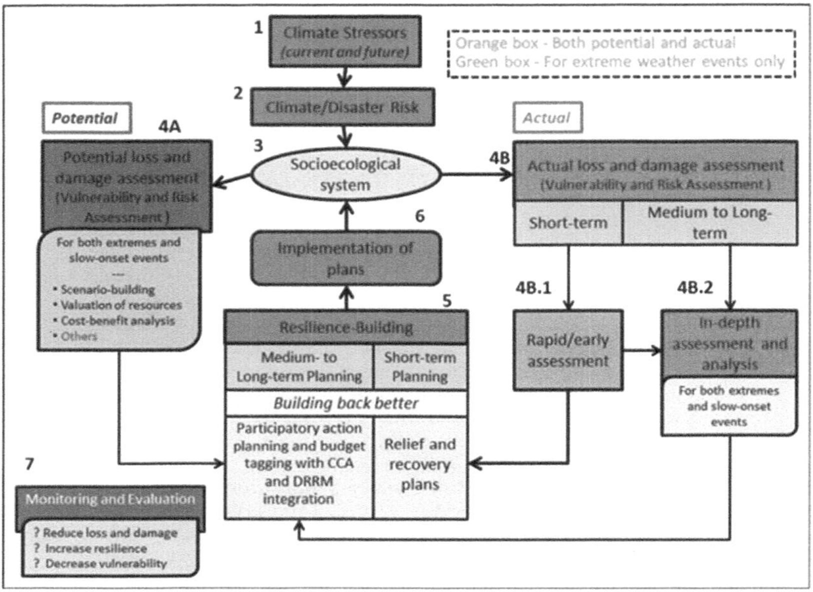

Fig. 1 Regional L&D integration framework

- The current and potential climate stressors to social ecological systems are presented by the linkage from blocks 1–3. According to Warner et al. (2013), the change and uncertainty of climate variabilities in ecosystems are climate stressors. The stressors would impact the magnitude of the disaster and socioecological systems, in which the socioecological system is a complex system linking social actors and community institutions. They would interact and change to adapt to the change in the environment (including climate factors). As a result of their adaptation strategies and interactions, the impacts of climate stressors in the systems would vary.
- Block 4 L&D assessment

  + Under a proactive approach, block 4A shows further L&D information for a better DRR and CCA activity for both slow- and sudden-onset events. To obtain the necessary information, L&D assessment activities have taken into account the potential socioeconomic impacts of climate change. The assessment should

socioeconomic modelling, cost–benefit analysis and climate projections.

+ Block 4B presents current L&D assessments, categorised into both rapid assessment (4B.1) and in-depth assessment (4B.2). Data from 4B.1 could be used for quick relief (and recovery planing. Meanwhile, in-depth assessment (4B.2) should be applied for both sudden and slow-onset events, employing exiting national assessment tools and specialized research from assessment teams. The results would contribute to developing comprehensive planning for short, medium and long-term DRR and CCA activities, thereby reducing maladptation to climate change and enhancing the resiliene of affected areas. Results of 4B.2 would also be used to define lessons learned after extreme events. The in-depth assessments would give more specific information on intangible L&D, which rapid assessment would not provide.

- Block 5—Resilience building (improving the resilience of socioeconomic and ecological systems) and Block 6—implementation.

  + The information from block 4 is a significant input for attempts to enhance resilience of the system as well as the effectiveness of DRR &CCA activities. To improve the resilience of the system, managers need to consider short- medium- and long-term plans. Additionally, these assessments aim to address questions such as whether current policies may lead to maladaptation in the future, and if there is a necessity to establish new policies regarding budget allocation and strategies for DRR and CCA activities.

  + At block 6, managers might consider what measure should be taken to improve the implementation of activities. They should consider who needs to be involved and what steps are necessary to implement further activities that contribute to improving the resilience of the system. Additionally they should assess whether the resilience strategies proposed in include on the priority lists of the government.

- Block 7 highlights the significance of monitoring and evaluating the Integration Framework which links DRR, CAA and L&D. Sets of indicators should be established to monitor whether the issues of current systems are being addressed. Within the integrated framework, managers might consider if activities are effectively reducing L&D of the system; increase resilience of the system and if they help to reduce the vulnerability of socioeconomic and ecological systems to climate change impacts.

3.2.2 *SWOC Analysis of the Proposed Framework*

In 2017, Lasco et al. have pointed out 7 mains challenges to linkage DRR, CCA and L&D in the report of the APN project on Linkage between Climate Change Adaptation (CCA), Disaster Risk Reduction (DRR), and Loss and Damage (L&D): Case Studies in the Low-lying Coastal Cities of Indonesia, Philippines, Thailand and Vietnam. Those challenges are data, governance, translation, funding, assessment information, partnership and technical capacity.

The key-informal interview updated by additional research in 2022 is trying to fill up the SWOC analysis of the mentioned challenges to link DRR, CCA and L&D in the contexts of DRR and CCA policy changes in researched countries as well as the updated information from COP 26 and 27. Table 2 shows results of SWOC analysis of the holistic framework to linkage DRR, CCA and L&D.

4 Recommendation and Conclusion

- The research attempted to understand emerging issues of CCA and DRR activities with regard to L&D at the local, national and regional scales. It is clear that L&D is a new concept to many stakeholders who take part in DRR and CCA activities. However, the linkage between DRR, CCA and L&D could be necessary to increase the resilience to climate change and disaster risk in the researched countries. A holistic framework to link DRR, CCA and L&D is necessary for climate mainstreaming into socioeconomic development and to raise early warning information for early action to deal with climate-related risks.
- There are existing frameworks for L&D assessments in the countries studied. Information from the framework is mostly economic costs of tangible loss. There is a need to strengthen baseline data for intangible L&D. Additionally, due to a lack of available data, assessment information for specific sectors, risk governance, communication methods, technical capacities, a strong partnership and funding assistance, there are many difficulties to enhance the application of the proposed framework to integrate DRR, CCA and L&D.
- Under new conditions in 2023 following COP 27, where L&D has become an emerging topic, the mentioned difficulties to link DRR, CCA and L&D remain obstacles in the effectiveness of DRR and CCA management systems in the countries studies for several reasons: (1) L&D is a new concept to many stakeholders and there

Table 2 SWOC analysis of the holistic framework to link DRR, CCA, and L&D

	Strengths	*Weaknesses*	*Opportunities*	*Constraints*
Data	– There are several available data assessment systems relating to L&D – There are available forums and websites which attempt to integrate available information on L&D, DRR and CCA into maps	– Lack of standardized database – Lack of integrated data from difference sectors – Differences in data collection methods – Lack of available data – Most of available data are economic cost of tangible loss, injuries and death tools. There is a lack of data on intangible loss and damage	– Governments and international agencies are putting efforts into making data more readily available, including conducting more research on baseline data; governments are establishing more policy which defines standards and tools in data collection; integrating available data into the geographic information systems (GIS) for integrated management; general index such as climate resilience index are developing and testing in those countries	– L&D is still a new concept, there could be a misunderstanding among those who collecting and use the data – Social and intangible assessments are still a question mark to assess and integrate

(continued)

Table 2 (continued)

	Strengths	*Weaknesses*	*Opportunities*	*Constraints*
Gover-nance	– DRR and CCA activities are considered as important governance activities in the researched countries – There are existing DRR and CCA frameworks in the countries	– Weak implementation of institutional arrangements to conduct L&D initiatives – A shortage of capability among authorities from local to central level to well distribute resources – Integration of L&D, CCA and DRR is encouraged/not compulsory in	– Integration of L&D, CCA and DRR initiatives is encouraged in development planning – International donors/agencies are providing technical support for climate proofing/manstreaming in development activities/projects in those countries	– It is likely to have a lack of political will to implement existing policies related to L&D, CCA and DRR in the researched countries when they need to harmonize with the economic development – L&D is a very new concept, government agencies could find it hard to understand and mainstream the integration of these three interrelated concepts

	Strengths	*Weaknesses*	*Opportunities*	*Constraints*
Trans-lation/ Com-munication	– The disaster-risk communication system varies to enhance the effectiveness of the communication process, e.g. meetings and documents between national agencies and relevant stakeholders would help stakeholders to understand the potential impacts of climate change and disaster risks to their planning and implement appropriate DRR & CCA activities	– The media—a very important risk in communications lacks awareness of L&D issues as well as could not have enough materials to provide an effective communication process – Translating the scientific knowledge/ early warning information to meet the users' demand is still challenging especially for a new concept such as L&D. Translators might need to understand users needs in advance before converting the information on potential L&D into early warning information for DRR and CCA activities	– Understanding L&D and the linkages between DRR, CCA and L&D would also support the (disaster risk/ climate change) early warning system. The two topics of L&D and early warning systems were hot at UNFCCC COP 27. L&D could help experts understand better the demand of affected communities. Hence, there are opportunities to link 2 topics and support the better risk communications – Climate services is promoting for years, climate services. The system could be seen as a pilot for risk communication to integrate L&D, DRR and CCA – Thanks to the development of GIS systems, many information could be integrated into the system to provide understandable information to relevant stakeholders	– L&D is a very new concept and there is a need to improve the understanding of experts/communicators who help to interpret the information to users

(continued)

Table 2 (continued)

	Strengths	*Weaknesses*	*Opportunities*	*Constraints*
Funding	– There is annual state funding for DRR, CCA and L&D-related activities in the countries – Risk transfer merchandise (e.g. risk insurance is applied/ tested in the research countries)	Lack of financial assistance for L&D assessment as the in-depth L&D assessment would require experts from different sectors. The wonder of financial assistance is one of the main obstacles of L&D assessment in the researched countries	– L&D concepts are becoming more popular internationally and it was a topic at UNFCCC COP 27. There are opportunities for financial application to improve L&D assessment systems as well as to integrate L&D, DRR and CCA -Private companies, including the insurance sector, are becoming more interested in risk transfer	The appearance of pandemics (eg. Covid 19) would significantly require tremendous expense to control and, consequently, this would result in a lack of financial resources to invest in DRR and CCA activities
Assessment information (for specific sectors)	– Ministries in the countries are considering CCA and DRR in their development plans (e.g. in Vietnam, along with the National action plan on CCA, each ministry has to establish related CCA action plans)	– Lack of sector-specific data and a holistic framework to assess the risk of each sector. Consequently, policy/decision makers could find it hard to plan, create and implement risk management activities	– Support from international agencies to assess potential L&D in specific sectors – Along with state actors, there are more non-state actors (NGOs, academia, firms, etc.) taking part in DRR & CCA activities and they would support L&D activities and encourage linkages between L&D, CCA and DRR	– It requires funding and technical support – The assessment results are rarely published

	Strengths	*Weaknesses*	*Opportunities*	*Constraints*
Technical capacity	– Key stakeholders have been trained to conduct and interpet related L&D assessments	– Many government staffs would/policy and decision makers would find it hard to understand L&D and or conduct related assessments as they would continue to change their position and responsibilities in their agencies – The accuracy of the weather forecast and climate projection is improving. Related stakeholders would believe the warning information easier	– There are more programmes to strengthen the technical capacity (from assessment, forecasting, understanding, etc.) of those involved in DRR, CCA and L&D framework – Investing in climate change adaptation activities and green growth is a trend of green finance. Hence, more private investment would be allocated to improve the technical capacity of the proposed framework – WMO has developed the Global Climate Service Framework with clear guidelines for countries	– L&D is a very new concept to get attention from stakeholders – Investors would prefer more profitable sectors such as climate change mitigation activities

(continued)

Table 2 (continued)

	Strengths	*Weaknesses*	*Opportunities*	*Constraints*
Partner-ship	– Policy systems in the researched countries are encouraging the involvement of stakeholders (business sectors, NGOs, communities, academia, international agencies, etc.) into DRR, CCA and L&D systems (e.g. the four-on-the-spot motto in Vietnam) – Currently, along with state actors, non-state actors are taking part in the system more frequently	– There are still a lack of guidelines for the involvement of non-state actors in the system – Lack of a common forum for those stakeholders to communicate and share information	– Researched countries are strengthening international cooperation – International agencies are supporting stakeholder engagement in projects/development activities including DRR, CCA and L&D activities	– L&D is a very new concept and there is still a lack of awareness of several stakeholder in CCA, DRR and L&D activities

are limited opportunities to strengthen knowledge and technical capacities of stakeholders in the management systems; (2) there is a need to establish guidelines and regulations on L&D in the existing DRR and CCA framework in those countries; and (3) a lack of financial assistance is still one of the main obstacles. There is a need to strengthen the active involvement of stakeholders including non-state actors in the framework of linking DRR, CCA and L&D.

Acknowledgements The article was supported by the project "***Assessing the Linkages between Climate Change Adaptation (CCA), Disaster Risk Reduction (DRR), and Loss and Damage (L&D): Case Studies in the Low-lying Coastal Cities of Indonesia, Philippines, Thailand and Vietnam***" reference CAF2015-RR05-CMY-Lasco, funded by the Asia-Pacific Global Network for Global Change Research (APN).

References

Buhr, B., Volz, U., Donovan, C., Kling, G., Lo, Y. C., Murinde, V., & Pullin, N. (2018). Climate Change and the Cost of Capital in Developing Countries.

Dodman, D., Hayward, B., Pelling, M., Castan Broto, V., Chow, W., Chu, E., Dawson, R., et al. (2022). "Cities, Settlements and Key Infrastructure." In Climate Change 2022: Impacts, Adaptation, and Vulnerability. Contribution of Working Group II to the Sixth Assessment Report of the Intergovernmental Panel on Climate Change, edited by Pörtner H.-O., et al. Cambridge: Cambridge University Press.

Harmeling, S., & Eckstein, D. (2013). Global Climate Risk Index 2013.

Huq, S., Roberts, E., & Fenton, A. (2013). Loss and Damage. *Nature Climate Change, 3*(11), 947–949.

IPCC. (2007). Climate Change 2007: Synthesis Report. Contribution of Working Groups I, II and III to the Fourth Assessment Report of the Intergovernmental Panel on Climate Change.

IPCC. (2014). WGII Report: AR5 Climate Change 2014: Impacts, Adaptation, and Vulnerability.

IPCC. (2018). IPCC Special Report on the Impacts of Global Warming of 1.5°C.

IPCC. (2019). IPCC Special Report on Climate Change, Desertification, Land Degradation, Sustainable Land Management, Food Security, and Greenhouse Gas Fluxes in Terrestrial Ecosystems.

IPCC. (2022). Climate Change 2022: Impacts, Adaptation, and Vulnerability. Contribution of Working Group II to the Sixth Assessment Report of the Intergovernmental Panel on Climate Change, edited by H.-O. Pörtner, D.C. Roberts, M. Tignor, E.S. Poloczanska, K. Mintenbeck, A. Alegría, M.

Craig, S. Langsdorf, S. Löschke, V. Möller, A. Okem, B. Rama (p. 3056). Cambridge, UK and New York: Cambridge University Press. https://doi.org/10.1017/9781009325844

Jibiki, Y., Pelupessy, D., Sasaki, D., & Iuchi, K. (2020). "Implementation of Post Disaster Needs Assessment in Indonesia: Literature Review." *Journal of Disaster Research, 15*(7), 975–980.

Kreienkamp, J. U. L. I. A., & Vanhala, Lisa. (2017). "Climate Change Loss and Damage." *Global Governance Institute* (2017), 1–28.

Lassa, J. A. (2013). Disaster Policy Change in Indonesia 1930–2010: From Government to Governance? *International Journal of Mass Emergencies & Disasters, 31*(2), 130–159.

Lasco, R. D., Boer, R., Chinvanno, S., & Ninh N.H. (2017). Climate Adaptation Framework Regional Research Final Report: "Assessing the Linkages between Climate Change Adaptation (CCA), Disaster Risk Reduction (DRR), and Loss and Damage (L&D): Case Studies in the Low-lying Coastal Cities of Indonesia, Philippines, Thailand and Vietnam".

OECD. (2009). *Integrating Climate Change Adaptation into Development Co-operation*. Paris, France, viewed 10 Aug 2022, https://www.oecd.org/env/cc/44887764.pdf.

Shaw, Rajib, SVRK Prabhakar, and Yohei Chiba. "SDGs, DRR and CCA: Potential for Strengthening Inter-Linkages." Institute for Global Environmental Strategies, 2016. http://www.jstor.org/stable/resrep02906.

UNDRR. (2020). Disaster Risk Reduction in The Republic of Indonesia: Status Report 2020. Bangkok, Thailand, United Nations Office for Disaster Risk Reduction (UNDRR), Regional Office for Asia and the Pacific.

UN ECLAC. (2014). Regional Perspectives on Sustainable Development: Advancing Integration of its Three Dimensions Through Regional Action.

UNFCCC. (2008). Report of the Conference of the Parties on its thirteenth session, held in Bali from 3 to 15 December 2007. In Addendum. Part Two: Action taken by the Conference of the Parties at its thirteenth session Decisions adopted by the Conference of the Parties.

UNFCCC. (2013). Technical Paper: Non-economic Losses in the Context of the Work Programme on Loss and Damage.

UNFCCC. (2014). Report of the Conference of the Parties on Its Nineteenth Session, held in Warsaw from 11 to 23 November 2013, Proceedings. Part One. UN.

UNFCCC. (2016) Report of the Conference of the Parties on Its Twenty-First Session, held in Paris from 30 November to 13 December 2015.

UNISDR. (2017). UNISDR Annual Report 2017: 2016–17 Biennium Work Programme Final Report.

Warner, K., & K. Van der Geest. (2013). Loss and Damage from Climate Change: Local-level Evidence from Nine Vulnerable Countries. *International Journal of Global Warming, 5*(4), 367–386.

CHAPTER 11

Linkages Between Disaster Risk Reduction and Climate Change Adaptation in the Context of Increasing Climate Change-Induced Loss and Damage in Vietnam

Le Minh Nhat and Dang Quang Thinh

Abstract Recently, worldwide loss and damage caused by sudden extreme weather and climatic slow-onset events have increased dramatically and adversely affected the development of many countries. Vietnam is no exception to this. In other words, climate change-induced losses and damages in Vietnam are inevitable and tend to increase in the future.

L. M. Nhat (✉)
Department Natural Disaster Response, Vietnam Disaster Management Authority (VNDMA), Hanoi, Vietnam
e-mail: nhatkyoto@gmail.com

D. Q. Thinh
Climate Change Research Center, Vietnam Institute of Meteorology, Hydrology and Climate Change (IMHEN), Hanoi, Vietnam

V. Wijenayake et al. (eds.), *Linking Climate Change Adaptation, Disaster Risk Reduction, and Loss & Damage*,
https://doi.org/10.1007/978-981-99-8055-0_11

Therefore, developing a mechanism to address the loss and damage at the national level through strengthening the linkage between climate change adaptation and disaster risk reduction is of great importance. This paper aims to derive potential linkages between adaptation and disaster risk reduction from the perspective of loss and damage through legal and policy analysis and expert interviews. An important component of both disaster risk management and climate change adaptation is the appropriate allocation of efforts in disaster management, disaster risk reduction and risk sharing through the risk management framework.

Keywords Climate change · Adaptation · Disaster risk reduction · Loss and damage · Mechanism · Vietnam

Highlights

- To ensure the success of an integrated mechanism for DRR and CCA to address L&D, a fundamental change in the thinking of those responsible for building and performing these activities is urgently needed.
- The integrated mechanism to address loss and damage requires a fundamental transition in policy formulation and implementation, from a sector-by-sector approach to a comprehensive, multi-sectoral approach.
- A national mechanism for integrating DRR, CCA for addressing L&D should also cover slow-onset events and non-economic losses apart from sudden-onset events and economic losses.

1 Introduction

In recent years, the loss and damage (L&D) caused by both sudden extreme climate events and slow-onset events have increased significantly, causing severe consequences and seriously affecting the development of countries worldwide including Vietnam. The L&D caused by climate change in Vietnam is significant and tends to increase. From 2006 to 2020, the frequency and intensity of natural disasters have increased in the context of climate change, which causes negative damage to people and properties, leading to 5304 deaths and huge property damage with an estimation of US$950 million, accounting for 0.65% of GDP per year (Figs. 1 and 2). The cost of adaptation, not including all long-term loss and damage impacts, is expected to reach 3–5% of GDP by 2030 (UNDP 2018).

The issue of L&D is attracting increasing attention from the international community and is one of the controversial topics at the Conference of the Parties (COP) to the United Nations Framework Convention on climate change (UNFCCC). At COP16, the Cancun Agreement emphasised reducing L&D caused by climate change through increased international cooperation and capacity building. Accordingly, the L&D Program focuses on three areas: (1) Risk assessment of L&D due to climate change impacts; (2) Identifying approaches to address climate change damage (including the impact of extreme weather events and slow-onset events); and (3) Identifying the role of the Climate Convention in assisting in addressing loss-related issues.

The concept of L&D emerged for the first time during the drafting of the UNFCCC in 1991 and L&D was discussed within the UNFCCC for many years, but was only institutionalised in 2013 during COP19 with

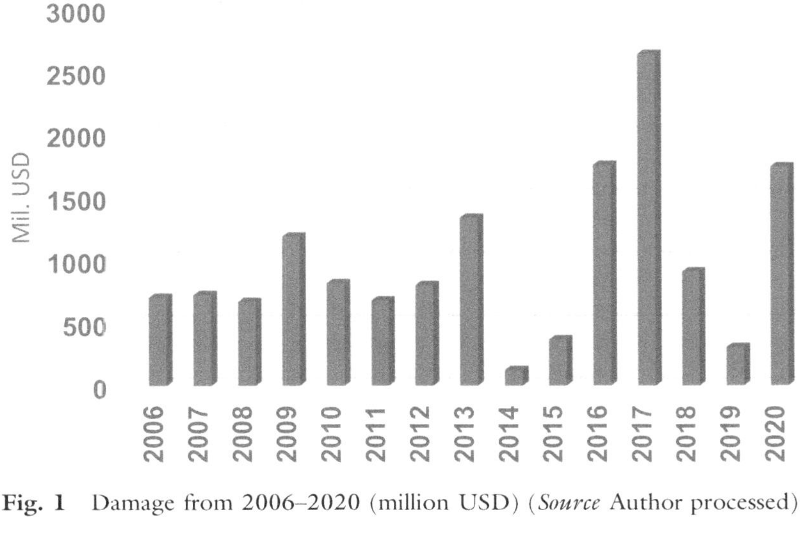

Fig. 1 Damage from 2006–2020 (million USD) (*Source* Author processed)

Fig. 2 Losses from 2006–2020 compared to GDP (%) (*Source* Author processed)

the establishment of the Warsaw International Mechanism for Loss and Damage (Jensen and Jabczyńska 2022). At COP21, the issue of loss and damage was embedded in the Paris Agreement. At COP25 in 2019, the UNFCCC parties established the Santiago Network on Loss and Damage to offer developing countries technical assistance on L&D. At COP26, the Glasgow Dialogue on funding for loss and damage was established. A breakthrough agreement was obtained at COP27 to provide "loss and damage" funding for vulnerable countries severely impacted by climate disasters (UNFCCC 2022). However, terms, who will pay, and who will receive funds have not been made clear, but the "transitional committee" was established to discuss how to operationalise both the new funding arrangements and the fund at COP28.

So far, there has not been an internationally common consensus on the main issues related to L&D. However, in order to maximise the effectiveness of the international mechanism, each country also needs to develop a separate L&D settlement mechanism.

One of the solutions to develop a mechanism to address L&D at the national level is to strengthen the linkage between climate change adaptation (CCA) and disaster risk reduction (DRR). Considering both theoretical perspective and practical implementation experiences in Vietnam, the fields of DRR and CCA have many opportunities for an effective coordination. However, connectivity enhancement faces many difficulties and challenges due to scale, nature and implementation time differences. This paper focuses on discussing the following topics: (1) The status of integration of DRR and CCA in Vietnam in terms of policy and planning tools; (2) Integrating approaches of DRR and CCA to address loss and damage; and (3) Loss and damage in Vietnam.

2 Methodology

This study has been built on a comprehensive internal review of a large number of policies, strategies and planning documents for DRR, CCA and L&D. Reviewed documents for the desk research included Law on Natural Disaster Prevention, National Climate Change Strategy (NCCS), Nationally Determined Contribution (NDC), National Adaptation Plan (NAP), Plan for Implementation of the Paris Agreement (PIPA), national climate change policies and relevant ministerial and sectoral policies. Evaluation of secondary data from existing research and reports and agencies inside and outside of Vietnam was also carried out.

3 Results and Discussion

3.1 Similarities and Differences Between Loss and Damage—Climate Change Adaptation—Disaster Risk Reduction

Till now, the UNFCCC has no official definition of L&D. Warner, K (Warner & Van der Geest 2013) defined loss and damage as an adverse consequence of climate variability and climate change that people have not been able to deal with or adapt to. This comprises the inability to respond to climate stressors (drought, floods, sea level rise, etc.) and the costs associated with the adaptation measures. Moreover, costs and adverse effects can be both economic and non-economic. So far, the definition largely accepted and popularly used in the literature defines L&D as "impacts of climate change that have not been, or cannot be, avoided through mitigation or adaptation efforts". L&D is also related to mitigation because the potential costs of future climate change response depend on mitigation efforts on a global scale.

Disaster risk is the damage caused by natural disasters in a certain period of time (UNISDR 2009a). Disaster risk refers to the likelihood of natural disasters rather than describing actual disaster events.

The disaster risk is defined as a reflection of the concept of disaster resulting from continuing present risk conditions. Disaster risk includes different types of the potential damage that are not easy to quantify (Twigg 2004). A hazard can only lead to a disaster if individuals and social systems are vulnerable to the effects of the hazard (UNISDR 2009b).

Climate change adaptation (CCA) is the adjustment of natural or human systems in response to actual or expected changes in the climate, to mitigate harm or to take advantage of opportunities to benefit from climate change (IPCC 2007). CCA generally refers to response to changes in the long-term trends of the climate and to environmental changes caused by climate. The term usually does not refer to short-term "adjustments" in response to short-term changes in climate. However, the difference between the two forms of short-term and long-term correction is not very clear. In fact, adaptation can be gradual as people and organisations make many short-term adjustments. In fact, the relationship between mitigation, CCA and L&D is viewed differently by the parties and differences between those concepts are open to interpretation (Jensen and Jabczyńska 2022). According to the Law on Hydrometeorology of Vietnam (2020), a natural disaster is an abnormal natural phenomenon that can cause damage to people, property, the environment, living conditions and socio-economic activities. Following the Law on natural disaster prevention and control (2013; amending and supplementing in 2020), natural disaster includes storms, tropical depressions, cyclones, heavy rain, flood, flash flood, inundation, landslide and land subsidence due to rain or flood, sea level rise, saline intrusion, heat, drought, harmful cold, hail, hoarfrost, earthquake and tsunami.

There is a difference between climate change adaptation and disaster risk reduction (Fig. 3). CCA looks at gradual effects of climate change (slow-onset events e.g. sea level rise, air temperature increase, snow melt, desertification…) while DRR also considers non-climate-related hazards e.g. earthquakes, volcanic eruption…

By contrast, DRR and CCA also have the considerable convergence as they both have common concerns about increased frequency and/or intensity of climate-related hazards such as floods, cyclones, heatwaves, droughts and landslides. Table 1 shows conceptual convergence between DRR and CCA.

Loss of damage becomes part of CCA programs, especially the Cancun Adaptation Framework. Countries should develop their adaptation plans taking into account the inevitable losses and damages caused by climate

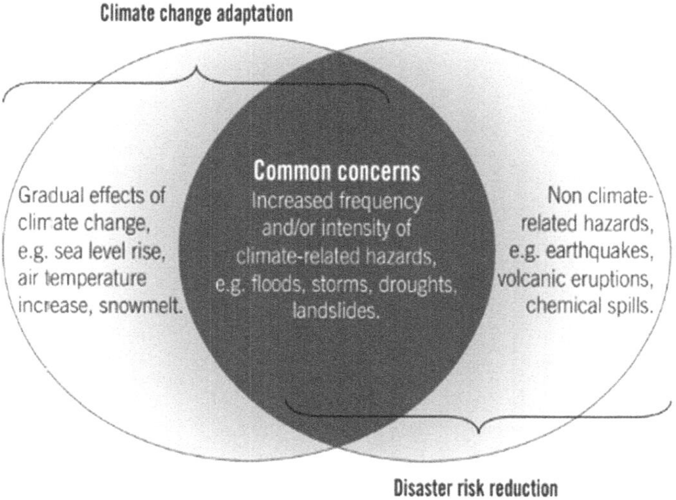

Fig. 3 Difference between climate change adaptation and disaster risk reduction (*Source* Turnbull et al. 2013)

change. Distinguishing adaptive action from loss-resolving action depends heavily on determining adaptive capacity limits. Many countries have difficulty identifying the limits of adaptive capacity and it is recommended that an adaptation framework be developed that clearly outlines the adaptive capacity of the community.

Table 1 Terminologies of DRR and CCA

Characteristic	*Disaster risk reduction*	*Climate change adaptation*
Scope	All hazards. Technological hazards and natural hazards	Climate-related hazards, extreme weather events and slow onsets
Temporality	Short-term	Long-term
Hazard	A process, phenomenon or human activity that may cause loss of life, injury or other health impacts, property damage, social and economic disruption or environmental degradation	The potential occurrence of a natural or human-induced physical event or trend that may cause loss of life, injury, or other health impacts, as well as damage and loss to property, infrastructure, livelihoods, service provision, ecosystems and environmental resources
Exposure	The situation of people, infrastructure, housing, production capacities and other tangible human assets located in hazard-prone areas	The presence of people; livelihoods; species or ecosystems; environmental functions, services, and resources; infrastructure; or economic, social, or cultural assets in places and settings that could be adversely affected

Characteristic	*Disaster risk reduction*	*Climate change adaptation*
Vulnerability	The conditions determined by physical, social, economic and environmental factors or processes which increase the susceptibility of an individual, a community, assets or systems to the impacts of hazards	The propensity or predisposition to be adversely affected. Vulnerability encompasses a variety of concepts and elements including sensitivity or susceptibility to harm and lack of capacity to cope and adapt
Risk	The potential loss of life, injury, or destroyed or damaged assets which could occur to a system, society or community in a specific period of time, is determined probabilistically as a function of hazard, exposure, vulnerability and capacity	The potential for adverse consequences where something of value is at stake and where the occurrence and degree of an outcome are uncertain. In the context of the assessment of climate impacts, the term risk is often used to refer to the potential for adverse consequences of a climate-related hazard, or of adaptation or mitigation responses to such a hazard, on lives, livelihoods, health and well-being, ecosystems and species, economic, social and cultural assets, services (including ecosystem services), and infrastructure. Risk results from the interaction of vulnerability (of the affected system), its exposure over time (to the hazard), as well as the (climate-related) hazard and the likelihood of its occurrence

Source Authors complied based on UNDRR (2016, 2020) and IPCC (2018)

3.2 *Integrating the Approach of Disaster Risk Reduction with Climate Change Adaptation to Address Loss and Damage*

This integration presents challenges in terms of time and space scales, benchmarks, knowledge and experience (Birkmann and Teichman 2010).

Time and Space Scale: DRR is concentrated at the local level where the disaster is directly impacted while addressing the impact of climate change through adaptation measures mainly of interest at the national and global levels (Schipper 2009). Local government plays an important role in managing and operating DRR activities with the active participation of local people (Buckle et al. 2010). Strategies and plans addressing climate change impacts are mainly developed at the national level and often consider long-term goals. Case studies in Vietnam indicated that national policy directions are only partially implemented due to a significant lack of technical and financial resources, capacity building and an integrated institutional scheme dedicated to climate change. Short-term reactive responses often override proactive and preventive strategies, while only a few actions address structural drivers of vulnerability (Espagne et al. 2021).

In addition, there is a substantial difference in the time scale between DRR and CCA (Adger 2009). Disaster risk activities often focus on emergency relief activities, occurring quickly, with limited attention to long-term prevention and proactive response activities. The main reason could be the fact that funding for implementing disaster response activities is only provided for a short period of time (Birkmann 2010). From a practical perspective, this approach has limitations such as a lack of initiative and underestimating the importance of developing a CCA strategy and long-term DRR to build resilience to long-term climate change impacts (Oliver-Smith 2013).

The final limitation is the lack of rationality in the coordination of issues related to CCA and DRR at the national level (Maplecroft 2010). In Vietnam, the task of DRR is often tied to the Ministry of Agriculture and Rural development (MARD)'s responsibilities, while climate change's responsibility is assigned to the Ministry of Natural Resources and Environment (MONRE).

In Vietnam, many policy actions to respond to climate change for specific sectors and localities have been developed recently. However, at present, a number of programs, master plans and plans for climate change response are not guaranteed to be inter-regional and inter-sectoral. For example, the sharing of water resources; planning the construction of hydropower projects and the demand for irrigation and water use in daily life and agricultural production; planning for flood drainage and coastal urban development have not yet ensured inter-regional and inter-sectoral characteristics; building the sea dyke program has not been connected with traffic planning; developing residential clusters according to the green urban model, green architecture, etc. Therefore, it is necessary to develop an integrated master's adaptation program of DRR and CCA based on inter-regional and inter-sectoral loss and damage settlement.

Evaluation Criteria: Currently, there are no official standards or criteria to guide the implementation and monitoring of adaptation activities (Birkmann and Teichman 2010). In post-disaster recovery, affected structures are often rebuilt and restored to their old status, instead of building and improving infrastructure in the direction of considering the potential impacts of climate change and taking into account adaptation (De Soto 2010). Another conceptual challenge is disagreement on two areas, while some argue that CCA must be integrated into DRR, others argue that DRR should be considered as a cross-cutting theme in CCA (Birkmann and Teichman 2010).

In Vietnam, the M&E system for adaptation activities at the national level was promulgated in early 2022. This provides the basis for managing, coordinating and improving the effectiveness of CCA activities and state management of climate change.

Knowledge and Experience: In practice, information about climate change and extreme weather events has not been effectively communicated to local communities (Gaillard and Mercer 2013). On the contrary, the experience that has been synthesized and accumulated during thousands of experience years of local disaster response has not been effectively utilized and integrated into strategic planning for adaptation and disaster risk reduction for the long term.

One major difficulty facing the linkage between climate change and disaster risk is the uncertainty about the extent of the impacts of climate change on the increasing occurrence and intensity of extreme weather events such as storms, floods, droughts, etc. (Schipper 2009).

3.3 Loss and Damage in Vietnam

In Vietnam, the Ministry of Agriculture and Rural Development is the focal point for DRR while the Ministry of Natural Resources and Environment is responsible for coordinating and guiding the implementation of activities to respond to climate change. Over the past years, Vietnam has institutionalised international commitments on climate change-associated loss and damages. Assessment of loss and damages has been stipulated in the Law on Environmental Protection (2020) and specified by Circular No. 01/2022/TT-BTNMT. The Ministry of Agriculture and Rural Development and the Ministry of Planning and Investment have issued the Joint Circular No. 43/2015/TTLT-BNNPTNT-BKHĐT on guiding the making statistics and assessment of the damage caused by natural disasters.

Line ministries have coordinated a number of activities to maximize the role of policies, programs and projects on DRR and CCA. Especially, the National Forum on DRR and CCA is held annually with the chair of the Ministry of Agriculture and Rural Development. This forum represents the close cooperation between the Ministry of Agriculture and Rural Development, the Ministry of Natural Resources and Environment and other stakeholders in sharing information, experiences, insights and encouraging discussion to select policy choices for DRR and CCA (Table 2).

The integration and coordination between the two areas is a responsibility placed on the National Steering Committee on Natural Disaster Prevention and Control and the National Committee on Climate Change, whereby each Board is responsible for developing a management plan, promoting inter-sectoral and multi-sectoral coordination. DRR activities need to be integrated with CCA activities, and DRR strategies can contribute greatly to adaptation activities, especially the improvement of DRR activities, policy, institutional development, technical documentation and supporting tools (UNISDR 2009a).

To coordinate and develop an integrated mechanism for DRR and CCA to effectively deal with loss and damage, the National Committee on Climate Change and the Committee on Prevention and Control of Natural Disaster need to develop a coordination mechanism and provide specific guidance for stakeholders.

Examining the integration of Disaster Risk Reduction (DRR) with Climate Change Adaptation (CCA) from a loss-damage perspective

Table 2 Policies related to climate change adaptation and disaster risk reduction

Disaster risk reduction	*Climate change adaptation*
Hyogo Framework for Action (2005)	United Nations Framework Convention on Climate Change (1992)
National Report on Disaster Reduction in Vietnam prepared for the World Conference on Disaster Reduction, Kobe-Hyogo, Japan (2005)	The Initial National Communication to the United Nations Framework Convention on Climate Change (2003)
The ASEAN Agreement on Disaster Management and Emergency Response (AADMER) (2005)	The Second National Communication of Vietnam to the United Nations Framework Convention on Climate Change (2010)
Incheon Declaration Road-Map and Action plan (2010)	The Third National Communication of Vietnam to the United Nations Framework Convention on Climate Change (2019)
Law on Natural Disaster Prevention and Control (2013) and Law on amending and supplementing a number of articles of the Law on Natural Disaster Prevention and Control and the Law on Dikes (2020)	Law on Environmental Protection (2020)
Directive No. 42-CT/TW on strengthening the Party's leadership in Natural Disasters Management (2020)	The Law on HydroMeteorology (2018)
Resolution No. 76/NQ-CP on Natural Disaster Prevention and Control (2018)	Resolution No. 24/NQ-TW on proactively responding to climate change, strengthening natural resources management and environmental protection (2013)
National Strategy for Natural Disaster Prevention, Response and Mitigation (NSNDPRM) to 2020 (2007)	Conclusion No. 56-KL/TW on continuing the implementation of the Resolution of the 11th-tenure Central Committee's 7th Plenum on proactively adapting to climate change, stepping up the management of natural resources and protecting the environment (2019)
The provincial set of indicators for assessing natural disaster prevention and control (2020)	The 2011–2015 National Target Programme to Respond to Climate Change (2008)
The National Strategy on Natural Disaster Prevention and Control to 2030 with a vision toward 2050 (2021)	The National Strategy on Climate Change (2011)
National Plan for Natural Disaster Prevention and Control to 2025 (2022)	The 2012–2020 National Action Plan on Climate Change (2012)
Decree No. 66/2021/ND-CP supporting the Law on Disaster Prevention and Control (2021)	The Plan for Implementation of the Paris Agreement (PIPA) (2016)
The "Project for community awareness raising and community-based natural disaster risk management, with a vision to 2030" (2021)	The National Action Plan for the Implementation of the 2030 Agenda for Sustainable Development (2017)
	The National plan to adapt to climate change (NAP) for a period of 2021–2030, with a vision by 2050 (2020)
	The Scheme on tasks and solutions to implement the results of the 26th Conference of the Parties to the United Nations Framework Convention on Climate Change (COP26) (2022)
	The National Strategy on Climate Change for the period up to 2050 (2022)

Source Authors have synthesized policies

underscores the global responsibility delineated in the Hyogo Framework for Action and the Climate Convention (Mitchell et al. 2010). This responsibility is particularly pertinent for Vietnam, as a signatory to both international conventions, which has demonstrated proactive engagement in fulfilling its obligations under these frameworks.

In Vietnam, several policies related to loss and damage have been issued. The Ministry of Agriculture and Rural Development and the Ministry of Planning and Investment issued Circular 43/TTLT-BNNPTNT-BKHDT dated June 19, 2013 guiding statistics and assessment of damage caused by natural disasters. The Ministry of Natural Resources and Environment has issued Circular 01/2022/TT-BTNMT dated 07/01/2022 on instructions for the determination of losses and damages by climate change including both economic and non-economic losses.

Vietnam has developed a complete, separate institutional system for DRR and CCA. However, in the field of DRR, the system of legal documents and implementation is assessed to be more complete and stronger than the policies and plans for the CCA. The government has established the National Steering Committee for Natural Disaster Prevention and Control, the National Committee on Climate Change and the National Committee for Disaster Incident Response and Locating and Rescue at various levels (e.g. central, provincial, local) to strengthen management and coordination capacity to prevent and respond to natural disasters effectively; initially established local climate change response offices.

Although both CCA and DRR are focused on addressing loss and damage, a little difference in the capacity to deal with each type of disaster and its manifestations still exists. Currently, many policies and institutions coping with extreme climate events are in place, but there are not many policies to deal with damage caused by slow-onset events and factors affecting vulnerability that occur over a long period.

A national mechanism for loss and damage is needed to promote the implementation of an approach to address climate change-related loss and damage through the following activities: improving knowledge and understanding of global risk management approaches; addressing the loss and damage caused by climate change; strengthening cooperation between the parties on the assessment and implementation of loss and damage assessment methods; step up action and support financial resources, technology, capacity building to deal with loss and damage.

Table 3 Addressing loss and damage

	Non-economic loss and damage	*Economic loss and damage*
Both sudden-onset and slow-onset events	Recognition and repair of loss (whether or not accompanied by financial payments)	Planned relocation/ assisted migration
	Enabling access/safe visits to abandoned sited	Reskilling and alternative livelihoods provision
	Active remembrance	Compensation and other social protection measures
	Counselling	
	Official apologies	
Sudden-onset events		Short- and long-term recovery and rehabilitation
		Support for rebuilding livelihoods
		Insurance and risk transfer

Source Jensen and Jabczyńska (2022)

Besides economic loss and sudden-onset events, the L&D national mechanism should also cover slow-onset events and non-economic losses (Table 3).

To reduce the risk of loss and damage, developing guidelines requires a combination of CCA perspectives, approaches and institutions implementing DRR. The need to address the direct impacts of extreme climate events and the long-term effects of climate change with a slow-onset underscores the importance of integrating DRR to climate change adaptation to address losses and damages.

4 Conclusion

Bridging the gap between short-term and long-term development requires fundamental funding allocation and use reforms. For DRR, long-term plans should be applied to ensure resilience . Meanwhile, the CCA plan needs to be implemented with specific activities, and a monitoring and evaluation system.

It is necessary to develop and consolidate a monitoring and evaluation system to ensure the synchronization between socio-economic development orientations with DRR and CCA activities. Regular assessment of risks and vulnerabilities will ensure that fluctuations are detected in time so that short-term goals can be replaced with long-term adaptation strategies. In addition, through data detailing and vulnerability assessment, monitoring and evaluation activities should pay more attention to identifying DRR activities that promote adaptive capacity and climate change resilience.

The integration of CCA and DRR should be considered in international agendas. In addition, the need for finance, technical assistance and capacity building should also be considered. This mechanism requires a fundamental policy formulation and implementation transition, from a sector-by-sector approach to a comprehensive, multi-sectoral one. In parallel with the above recommendations, the following steps should be taken:

- Ensure the allocation of dedicated funds from domestic and foreign financial sources for approved activities;
- Form a network of stakeholders working in DRR and CCA at the national and local levels, including vulnerable communities, to share knowledge and experience to deal with loss and damage;
- Develop a joint program to guide planning and coordination among stakeholders to implement DRR and CCA.
- Develop a mechanism for sharing responsibilities and powers in planning and developing CCA and DRR activities at all levels.

Finally, to ensure the success of an integrated mechanism for DRR and CCA, there must be a fundamental change in the thinking of those responsible for building and performing these activities. In addition, the Ministry of Agriculture and Rural Development and the Ministry of Natural Resources and Environment will continue to influence policy direction on DRR and CCA so there is a need to increase and strengthen coordination and exchange of information more often between the two ministries in years to come.

Acknowledgements The article was supported by the project "*Strengthening the capacity of the Office of the National Committee and focal points at the central and local levels on loss and damage through Disaster Risk Reduction and Climate Change Adaptation*" code CAF2014-CD02NMY-Nhat, funded by the Asia-Pacific Global Affairs Network (APN).

References

Adger, W. Neil. (2009). Social Capital, Collective Action, and Adaptation to Climate Change. *Economic Geography* 79(4): 387–404. https://doi.org/10.1111/j.1944-8287.2003.tb00220.x.

Birkmann, J. (2010). Global Response and Reconstruction: Stabilization Versus Destabilization—Challenges of the Global Disaster Response to Reduce Vulnerability and Risk Following Disasters. In A. Doelemeyer, J. Zimmer, and G. Tetzlaff (eds.), *Risk and Planet Earth—Natural Hazards, Vulnerability, Integrated Adaptation Strategies*. Stuttgart, Germany: Schweizerbart.

Birkmann, J., and von Teichman, K. (2010). Integrating Disaster Risk Reduction and Climate Change Adaptation: Key Challenges—Scales, Knowledge and Norms. *Sustainability Science* 5(2): 171–184.

Buckle, P., Birkmann, J., Renaud, F., Setiadi, N., Born, B. T., and Sunarto. (2010). Introduction: Challenges in DRR and CCA in South and Southeast Asia. In N. Setiadi, J. Birkmann, and P. Buckle (eds.), *Disaster Risk Reduction and Climate Change Adaptation: Case Studies from South and Southeast Asia*. Bonn: UNU-EHS.

De Soto, H. (2010). Rebuilding after Disasters. In G. Lizarralde, C. Johnson, and C. H. Davidson (eds.), *Rebuilding After Disasters: From Emergency to Sustainability*. London: Taylor and Francis.

Espagne, E., Nguyen, T. T. H., Tran, T. A. D., Rieber, A., Hoai, D. T. T., Mania, E., ... & Yen Ha, B. (2021). The Macroeconomics of Climate Change and Adaptation in Viet Nam. In E. Espagne (Ed.), *Climate Change in Viet Nam Impacts and Adaptation*.

Gaillard, J. C., and Mercer, J. (2013). From Knowledge to Action: Bridging Gaps in Disaster Risk Reduction. *Progress in Human Geography* 37(1): 93–114.

Intergovernmental Panel on Climate Change (IPCC). (2007). Climate Change 2007: Impacts, Adaptation and Vulnerability. In M. L. Parry, O. F. Canziani, J. P. Palutikof, P. J. van der Linden, & C. E. Hanson (Eds.), *Contribution of Working Group II to the Fourth Assessment Report of the Intergovernmental Panel on Climate Change*. Cambridge University Press.

Intergovernmental Panel on Climate Change (IPCC). (2018). Special Report on Global Warming of 1.5°C. Retrieved from https://www.ipcc.ch/site/assets/uploads/2023/06/SR15_Citation.pdf

Jensen, L., and Jabczyńska, P. (2022). Understanding Loss and Damage Addressing the Unavoidable Impacts of Climate Change, July 2022. https://www.europarl.europa.eu/RegData/etudes/BRIE/2022/733598/EPRS_BRI(2022)733598_EN.pdf.

Maplecroft. (2010). Big Economies of the Future—Bangladesh, India, Philippines, Vietnam and Pakistan—Most at Risk from Climate Change.

Ministry of Natural Resources and Environment. (2019). *The Third National Communication of Vietnam to the United Nations Framework Convention on Climate Change*. Vietnam Publishing House of Natural Resources, Environment and Cartography.

Mitchell, T., Van Aalst, M., and Villanueva, P. S. (2010). Assessing Progress on Integrating Disaster Risk.

National Assembly. (2013). Law on Natural Disaster Prevention and Control. *Law No. 33/2013/QH13* (issued on 06/19/2013).

National Assembly. (2020). Law on Environmental Protection. *Law No. 72/2020/QH14* (issued on 11/17/2020).

Oliver-Smith, A. (2013). Disaster Risk Reduction and Climate Change Adaptation: The View from Applied Anthropology. *Human Organization* 72(4): 275–282.

Prime Minister. (2008). *The 2011–2015 National Target Programme to Respond to Climate Change*. Decision No. 158/2008/QD-TTg (issued on 12/2/2008).

Prime Minister. (2012). *The 2012–2020 National Action Plan on Climate Change*. Decision No. 1474 (issued on 11/5/2012).

Prime Minister. (2020). *The National Plan to Adapt to Climate Change for a Period of 2021–2030, with a Vision by 2050*. Decision No. 1055/QD-TTg (issued on 07/20/2020).

Prime Minister. (2022). *National Strategy on Climate Change for the period up to 2050*. Decision No. 896/QD-TTg (issued on 07/26/2022).

Schipper, L. (2009). Meeting at the Crossroads? Exploring the Linkages Between Climate Change Adaptation and Disaster Risk Reduction. *Climate and Development* 1(1): 16-30.

Turnbull, M., Sterrett, C. L., and Hilleboe, A. (2013). *Towards Resilience: A Guide to Disaster Risk Reduction and Climate Change Adaptation*. Warwickshire, UK: Practical Action Publishing.

Twigg, J. (2004). *Disaster Risk Reduction: Mitigation and Preparedness in Development and Emergency Programming*. Overseas Development Institute (ODI).

UNDP. (2018). Issue Brief Climate Change Loss and Damage in Vietnam Contribution to the Review and Update NDC1.

UNDRR. (2016). *Report of the Open-ended Intergovernmental Expert Working Group on Indicators and Terminology*. Retrieved from https://www.undrr.org/publication/report-open-ended-intergovernmental-expert-working-group-indicators-and-terminology

UNDRR. (2020). *Disaster Risk Reduction and Climate Change Adaptation, Pathways for policy coherence in Sub-Saharan Africa*. Intergovernmental Panel on Climate Change (IPCC). Special Report on Global Warming of 1.5°C. Retrieved from https://www.ipcc.ch/site/assets/uploads/2023/06/SR15_Citation.pdf

United Nations International Strategy for Risk Reduction (UNISDR). (2009a). Adaptation to climate change by reducing disaster risk: Country practices and Lessons. *Briefing Note 2*.

United Nations International Strategy for Risk Reduction (UNISDR). (2009b). *UNISDR Terminology on Disaster Risk Reduction*.

UNFCCC. (2022). https://unfccc.int/news/cop27-reaches-breakthrough-agreement-on-new-loss-and-damage-fund-for-vulnerable-countries. Accessed on 28th November 2022.

Warner, K., and Van der Geest, K. (2013). Loss and Damage from Climate Change: Local-level Evidence from Nine Vulnerable Countries. *International Journal of Global Warming, 5*(4), 367–386.

CHAPTER 12

Policy Gaps and Needs Analysis for the Implementation of NDCs on Adaptation and Loss and Damage in Bangladesh, Nepal, and Sri Lanka

Dennis Mombauer and Vositha Wijenayake

Abstract This paper focuses on the Nationally Determined Contributions (NDCs) submitted by Bangladesh, Nepal, and Sri Lanka, as required by the Paris Agreement, which include actions related to both mitigation and adaptation and loss and damage. The paper examines the implementation gaps and needs in these countries for NDC components related to adaptation and loss and damage. The research uses legal and policy analysis, expert interviews, group consultations, and multi-stakeholder workshops to identify gaps and needs in technical, financial,

D. Mombauer (✉) · V. Wijenayake
SLYCAN Trust, Colombo, Sri Lanka
e-mail: dennis@slycantrust.org

V. Wijenayake
e-mail: vositha@slycantrust.org

V. Wijenayake et al. (eds.), *Linking Climate Change Adaptation, Disaster Risk Reduction, and Loss & Damage*,
https://doi.org/10.1007/978-981-99-8055-0_12

and institutional capacities, coordination, data access, research, knowledge management, gender-responsiveness, monitoring, reporting, and verification of actions. The paper recommends aligning and integrating NDCs with other related processes, such as National Adaptation Plans, Sustainable Development Goals, and the Sendai Framework on Disaster Risk Reduction, to build synergies. The paper also recommends developing data and knowledge-sharing mechanisms and common national monitoring and evaluation systems for NDC implementation processes to address the identified gaps.

Keywords NDCs · Adaptation · Loss and damage · Implementation · Climate action

Highlights

- Nationally Determined Contributions (NDCs) are at the heart of the Paris Agreement and play a key role for guiding and accelerating national and global climate action. They embody the principle of "common but differentiated responsibilities" and include not only commitments on climate change mitigation but for many developing countries—for example, Bangladesh, Nepal, and Sri Lanka—components related to adaptation and/or climate-induced loss and damage as well.
- A thorough and participatory gap analysis is necessary to identify existing gaps and needs for the effective implementation of these NDC components and build synergies with other relevant processes on climate change adaptation, sustainable development, and climate risk management.
- Key gaps identified include those related to human and institutional capacities (technical, financial, institutional, and others), coordination, data access, knowledge-sharing, research, MRV systems, and gender-responsiveness.
- Building capacities and strengthening international cooperation to close knowledge and capacity gaps could contribute to addressing needs related to the implementation of adaptation and loss and damage components of NDCs more effectively and successfully.

- There are important linkages and interconnections between climate change adaptation, disaster risk reduction, and loss and damage across the NDCs and NAPs of the three countries. Identifying potential synergies and complementarities in the existing institutional frameworks and capacities for implementation could facilitate a more efficient and successful utilization of resources and benefit both national and global goals.

1 Introduction

In 2015, 195 Parties to the United Nations Framework Convention on Climate Change (UNFCCC) negotiated and adopted the Paris Agreement, a legally binding international agreement to collectively work towards addressing climate change. The Paris Agreement aims to coordinate and strengthen global action to address the threat of climate change, taking into consideration the need for sustainable development and efforts to eradicate poverty, including by: UNFCCC (2015)

- Keeping global warming to well below 2 degrees Celsius above pre-industrial levels, and pursuing efforts to limit the temperature increase to 1.5 degrees Celsius above pre-industrial levels;
- Increasing the ability to adapt to climate change impacts and fostering resilience as well as low-carbon development;
- Mobilizing finance to achieve these goals and enable climate-resilient, low-emission development pathways.

A core concept of the Paris Agreement is known as "common but differentiated responsibilities," acknowledging the different degrees of responsibilities and capacities of individual Parties. This principle is embodied in the process of preparing, submitting, and implementing Nationally Determined Contributions (NDCs), which outline Parties' commitments for post-2020 climate action and are submitted every five years with increasing ambition, serving as a ratchet mechanism for both individual and collective climate action under the Paris Agreement. To date, all 195 Parties have submitted their first, second, or updated NDCs, which are recorded in a public registry maintained by the UNFCCC Secretariat (UNFCCC 2015).

The NDCs submitted by Parties are not only focused on mitigation but often contain sectors connected to climate change adaptation and loss and damage (L&D) as well, particularly for developing countries. Furthermore, NDC implementation in each country does not take place in isolation but is closely linked with other processes such as the National Adaptation Plan (NAP) process introduced under the Cancun Adaptation Framework (Decision 1/CP16, FCCC/CP/2010/7/Add.1), the Sustainable Development Goals (SDGs) under the 2030 Agenda for Sustainable Development, or the Sendai Framework for Disaster Risk Reduction 2015–2030 (UNFCCC 2021; Wijenayake 2019).

Bangladesh, Nepal, and Sri Lanka are three developing countries in the South Asian region that have included relevant components on adaptation and L&D in their original NDC submissions as well as their updated or second NDCs submitted in 2020/2021. However, as Parties to the UNFCCC and the Paris Agreement move from the formulation of NDCs towards implementation, it has become clear that it is challenging to turn NDC commitments into action, and that developing countries require support and climate finance to adequately and effectively implement their submitted NDCs for mitigation, adaptation, and L&D (SLYCAN Trust/APN 2019).

This research aims to identify key lessons learned from an analysis of policy gaps and needs for the implementation of adaptation and L&D components of all three NDCs. To achieve this, it is guided by the following main research questions:

1. To what extent, for which sectors, and in what ways do the NDCs of Bangladesh, Nepal, and Sri Lanka include and address adaptation and L&D?
2. What are the gaps and needs to implement these elements of the current NDCs related to capacities, institutions, policies, data, technology, and finance?
3. What are the entry points to address these gaps and needs on the national as well as regional level, align NDCs implementation with NAPs, SDGs, and other related processes, and harness synergies and co-benefits?

Additional objectives of the research are the sharing of knowledge and expertise related to addressing implementation gaps and needs;

enhanced regional cooperation for adaptation, addressing L&D, and implementing NDCs; and the development of multi-stakeholder-driven, gender-responsive, inclusive, intersectional, and participatory recommendations for addressing the identified gaps and needs in the three countries.

2 Methodology

This research has been prepared based on a comprehensive literature survey and an in-depth analysis of the NDCs of Bangladesh, Nepal, and Sri Lanka. Furthermore, it includes an analysis of the relevant policy and legal environment of the three countries. The desk research was interlinked with a consultative process that consisted of individual interviews with key stakeholders and a total of 18 consultations, three national workshops, and one regional workshop, engaging more than 350 stakeholders from different sectors and levels (SLYCAN Trust/APN 2019).

The workshops and consultations were conducted together with government entities under the NDC review process and focused on different NDC adaptation and L&D sectors. The process is described in more detail in the four country-level research papers on Sri Lanka, Bangladesh, and Nepal as well as the regional synthesis published as part of the original project (Wijenayake 2019).

As shown in Table 1, the reviewed documents for the desk research included National Adaptation Plans (NAPs), NAP Readiness Proposals, National Adaptation Programmes of Action (NAPAs), National Communications to the UNFCCC, national climate change policies, and relevant sectoral policies of all three countries. Preliminary findings were supplemented and validated through the multi-stakeholder consultation meetings and workshops as well as one-on-one interviews with key actors from the public sector and civil society.

Based on the initial analysis for the laws and policies related to climate change, disaster risk management, and sustainable development, sectoral meetings, workshops, and one-on-one interviews were conducted to validate preliminary findings for the NDC sectors related to adaptation and L&D. Key stakeholders from the government, civil society organizations (CSOs), research institutions, think tanks, the private sector, and academia were invited for the consultations. Data was collected through discussions with the stakeholders, their inputs on the present status of NDC implementation, and stakeholder input provided on gaps and needs for the implementation of the sectoral NDCs.

Table 1 Analysed policy and planning documents

Country	*Reviewed document*
Bangladesh	Initial National Communication to the UNFCCC (2002)
	National Adaptation Programme of Action (2005)
	Bangladesh Climate Change Strategy and Action Plan (2009/2022)
	Second National Communication to the UNFCCC (2012)
	National Disaster Management Policy (2015)
	Nationally Determined Contributions (2015/2021)
	GCF Adaptation Planning Support Readiness Proposal (2018)
	NDC Implementation Roadmap and Action Plan (2018)
	Third National Communication to the UNFCCC (2018)
	Bangladesh Climate Fiscal Framework (2020)
	National Strategy on Internal Displacement Management (2021)
	National Plan for Disaster Management 2021–2025 (2021)
	Mujib Climate Prosperity Plan 2022–2041 (2022)
	National Adaptation Plan 2023–2050 (2022)
Nepal	Initial National Communication to the UNFCCC (2004)
	National Adaptation Programme of Action (2010)
	Climate Change Policy (2011)
	Second National Communication to the UNFCCC (2014)
	Nationally Determined Contributions (2015/2020)
	GCF Adaptation Planning Support Readiness Proposal (2016)
	Third National Communication to the UNFCCC (2021)
	National Adaptation Plan 2021–2050 (2021)
Sri Lanka	Initial National Communication to the UNFCCC (2000)
	Second National Communication to the UNFCCC (2011)
	National Climate Change Adaptation Strategy 2011–2016 (2011)
	National Climate Change Policy (2012)
	Sri Lanka Comprehensive Disaster Management Programme 2014–2018 (2014)
	Nationally Determined Contributions (2015/2021)
	National Adaptation Plan for Climate Change Impacts in Sri Lanka 2016–2025 (2015)
	GCF Strengthen the Process and Capacity of Implementation of National Adaptation Plan of Sri Lanka Readiness Proposal (2020)
	National Agriculture Policy (2012, draft)
	Third National Communication to the UNFCCC (2022)
	Sri Lanka's Climate Prosperity Plan (2022)

Feedback and comments received from the meetings and workshops were incorporated into three national research papers and a regional comparative study based on the country papers, which this chapter uses as a foundation for further elaborating findings and updating them with inputs from the newly submitted second or updated NDCs in 2020–2021.

3 Results and Discussion

Bangladesh, Nepal, and Sri Lanka all have included components related to adaptation and L&D in their NDCs. As highly vulnerable developing countries, they have identified adaptation and addressing L&D as priority areas for implementation to protect human lives, livelihoods, wellbeing, and economic development (IPCC 2022).

However, the outlined commitments require finance, human and institutional capacities, access to technology, and other forms of support to move from commitments and plans to concrete and effective implementation. It is therefore crucial to identify the existing gaps and needs for implementation to enhance the enabling environment needed to mobilize means of implementation and catalyse action.

3.1 Nationally Determined Contributions

As stated above, the three selected countries have all incorporated commitments related to climate change adaptation and addressing L&D in their respective NDCs. Sri Lanka has a total of nine adaptation sectors as well as a dedicated chapter on L&D and one on means of implementation in its NDCs. Bangladesh's NDCs include several adaptation components as well as a section on disaster management and a separate chapter on implementation mechanisms. Finally, Nepal addresses eight thematic adaptation sectors (including disaster risk reduction and management) as well as four cross-cutting areas (including climate finance management) in its NDCs and commits to developing a national strategy and action plan on L&D by 2025. There are also other thematic aspects—such as human mobility—which are included to some extent in the NDCs and/or NAPs of all three countries and could be integrated across different sectors and stages of the NDC and NAP process (SLYCAN Trust 2022).

Table 2 NDC status and priority sectors

Country	*NDC status*	*Submission*	*Adaptation and L&D priority sectors*
Bangladesh	First NDC (updated)	26.08.2021	Agriculture and food security; water resources management; sustainable ecosystem and livelihood; disaster management
Nepal	Second NDC	08.12.2020	Agriculture and food security; water resources and energy; forests, biodiversity, and watershed conservation; health, drinking water, and sanitation; rural and urban settlements; industry, transport, and physical infrastructure; tourism, natural and cultural heritage; disaster risk reduction and management
Sri Lanka	First NDC (updated)	24.09.2021	Agriculture; fisheries; livestock; water; biodiversity; coastal and marine; health; urban planning and human settlement; tourism and recreation; loss and damage

(a) Priority sector in all three countries; **(b)** priority sector in two countries; **(c)** priority sector in one country

Table 2 shows the submission status for the three countries as well as the priority sectors identified in each NDC, which exhibit significant overlap.

All three countries have included priority sectors related to food systems, water resources, natural ecosystems, and disaster risk management or L&D; Nepal and Sri Lanka have both included human health and sanitation, human settlements, infrastructure and transport, and tourism and recreation; and only Sri Lanka has included a specific adaptation sector on coastal and marine areas.

For addressing climate-induced L&D, the NDCs contain the commitments shown in Table 3.

3.2 *National Adaptation Plans*

NAPs build on experiences and good practices from the National Adaptation Programmes of Action (NAPAs). They are voluntary submissions by developing countries and aim to reduce vulnerability to climate change impacts; build adaptive capacities and resilience; and strengthen the coherent integration of climate change adaptation into relevant new and existing policies, programmes, and activities (UNFCCC 2021).

The guiding principles for NAPs indicate that they should follow a country-driven, gender-sensitive, participatory, and transparent approach and take into consideration the needs of vulnerable groups, communities, and ecosystems. Guided by the best available science as well as

Table 3 L&D commitments by country

Country	*Commitments on L&D*
Bangladesh	• References the Pilot Program for Climate Resilience (PPCR), a targeted fund within the Climate Investment Funds, and other programmes and actions towards disaster management and emergency response
Nepal	• Highlights that Nepal aspires to avoid residual risks caused by L&D and to receive support (financial and otherwise) for risks that may still materialize • Commits to developing a national strategy and action plan on L&D by 2025 and ensuring that the national climate financing and investment framework will acknowledge specific sources of finance for L&D
Sri Lanka	• Commits to conducting a gap analysis to assess the current status and understanding of L&D • Commits to strengthening the existing weather and climate forecasting system • Commits to improving data management systems to record losses and damages per sector • Commits to establishing an overarching, nationally appropriate, functional institutional mechanism for L&D in line with the WIM • Commits to developing a comprehensive national risk management framework

traditional and indigenous knowledge, they should integrate and mainstream adaptation into relevant social, economic, and environmental policies and actions without being prescriptive or duplicating existing efforts (UNFCCC 2012).

Table 4 outlines the timeline and status of the NAP process for each of the three countries, starting with the formulation of their NAPs and culminating in the completed (Nepal, Sri Lanka) or ongoing (Bangladesh) development of a NAP.

3.3 Priority Sectors

There are considerable similarities between climate-related challenges faced in the region as well as response options and commitments included in the NDCs, which point to a greater usefulness of the gaps and needs assessment for implementation.

Table 4 Timeline of NAP process by country

Country	*Year*	*NAP process*
Bangladesh	2005	National Adaptation Programme of Action
	2009	Updated of NAPA
	2015	Roadmap for developing a NAP for Bangladesh
	2017	National stocktaking and revisiting of NAP roadmap
	2018/2019	GCF readiness proposal: Adaptation Planning Support for Bangladesh through UNDP
	2022	National Adaptation Plan (2023–2050)
Nepal	2006	National Adaptation Programme of Action
	2016	GCF readiness proposal: Adaptation Planning Support for Nepal through UNEP
	2021	National Adaptation Plan (2021–2050)
Sri Lanka	2006	National Action Programme for Adaptation to Climate Change
	2016	National Adaptation Plan for Climate Change Impacts in Sri Lanka 2016–2025 (NAP)
	2020	GCF readiness proposal: Strengthen the Process and Capacity of Implementation of National Adaptation Plan of Sri Lanka

The analysis of priority sectors for Bangladesh, Nepal, and Sri Lanka based on both NDCs and NAPs (Table 5) shows an even stronger alignment of adaptation priorities than the analysis of NDCs alone, with all three countries sharing six adaptation priority sectors and one sector related to L&D, with two countries each sharing an additional adaptation sector (tourism and recreation for Nepal and Sri Lanka, cross-cutting issues for Bangladesh and Nepal).

Common categories of commitments for actions on adaptation and L&D include mainstreaming adaptation and resilience into different sectors, reducing vulnerabilities, enhancing risk management, strengthening early warning and forecasting, diversifying livelihoods, improving practices and systems towards sustainability and more resource efficiency, developing risk transfer and insurance schemes, improving risk preparedness, strengthening safety nets and social protection systems, building capacities, creating awareness, and conducting risk and vulnerability assessments.

Table 5 Priority sectors by country

Sector	*Bangladesh*		*Nepal*		*Sri Lanka*	
	NDCs	*NAP*	*NDCs*	*NAP*	*NDCs*	*NAP*
Food systems						
Water resources						
Natural ecosystems and biodiversity						
Human health and sanitation						
Human settlements						
Risk management and loss and damage						
Infrastructure and transport						
Tourism and recreation						
Cross-cutting						
Coastal and marine						
Export agriculture						

3.4 *National Gaps and Needs*

Based on the priority sectors identified in NDCs and NAPs as well as a closer analysis of the three national policy and legal frameworks for climate action (including national climate policies and plans, mid- and long-term development strategies, disaster risk management plans, and legal or regulatory frameworks related to climate change and environmental protection), the following gaps and needs have been identified for each country individually as well as for the region (SLYCAN Trust/APN 2019):

3.4.1 Bangladesh

The Government of Bangladesh has submitted its first NDCs in 2015, approved an NDC roadmap in 2018, and submitted an updated version of the NDCs in 2021. The country has several effective, well-drafted policies and strategies to support the roadmap and NDC implementation across different sectors but has identified a need for further financial support on adaptation, including from multilateral agencies and the private sector. For this purpose, there is also a need to develop a more detailed monitoring system and remove barriers for private sector investment in adaptation (SLYCAN Trust/APN 2019).

The institutional arrangements for NDC implementation need to be strengthened through the clarification of mandates, responsibilities, and boundaries of different institutions related to adaptation planning and action. Gaps in coordination and exchange of information between different sectoral institutions and levels of government need to be closed, and climate change needs to be further mainstreamed into relevant sectoral policies.

Furthermore, relevant agencies and stakeholders need to be made more aware of the adaptation part of Bangladesh's NDCs, and capacities need to build within government ministries and line agencies to effectively coordinate, streamline, and implement actions related to NDC adaptation components.

L&D is addressed in the country through different means, including a social safety net programme, disaster management, relief, and emergency support schemes, rebate of agricultural loans, subsidies on agricultural input, and the planned establishment of a national mechanism on L&D. Scaling up climate finance for adaptation and L&D and further shifting risks and costs away from vulnerable frontline communities have been identified as a key need, as these costs are currently borne predominantly from the national budget and by frontline communities themselves (UNDRR 2015).

While the NDCs do not explicitly reference L&D and only contain components related to disaster management, the NAP establishes a direct link between adaptation, disaster risk reduction, and L&D. For example, the NAP specifies average losses and damages for different regions in the country, link this to adaptation and disaster risk reduction measures, and commit to introducing "innovation in reducing climate change and disaster risks and consequent losses and damages."

A detailed assessment of climate impacts and a stronger connection between the NDCs on disaster management and other actions related to L&D could potentially enhance access to support and funding. Gaps and needs in monitoring, evaluation, progress tracking, and reporting of NDC implementation and NAP progress should also be addressed to ensure effective, accountable, and transparent climate action in Bangladesh.

3.4.2 Nepal

The Government of Nepal has submitted its first NDCs in 2016 and second NDCs in 2020. Among the gaps and needs for the implementation of adaptation and L&D NDC components in the country, the participating stakeholders identified the need for improved institutional coordination as well as integrated actions. Further, capacity gaps and needs for the implementation of NDCs among key stakeholders have also been highlighted through research findings, as well as the need for available national-level finance to reach subnational and local-level institutions and stakeholders (SLYCAN Trust/APN 2019).

Additional gaps and needs that were highlighted through the research include a lack of explicit and trackable targets for L&D, the need for a robust MRV system that can track and report on implementation progress, and the need to address gaps in the tracking of climate finance, budget tagging, and access to finance for especially subnational and local-level actors.

In terms of synergies, the NDCs explicitly frame L&D in the context of residual risks and establish a link between (disaster) risk management and L&D while also aiming for a financial framework that addresses adaptation and L&D through the same framework but different sources of funding. Similarly, the NAP commits to enhancing policy coherence and a harmonized enabling environment for adaptation and disaster risk reduction (including through the Sendai Framework) while linking disaster risk reduction directly to aspects of L&D such as disaster displacement.

3.4.3 Sri Lanka

Sri Lanka submitted its first NDCs in 2016, followed by an updated submission in 2021, which included additional commitments for both mitigation and adaptation. Among the policy gaps and needs identified for implementation, the need for amendments of sectoral policies has been identified in the original research and has been partially addressed in the meantime, for example through the completed or ongoing development of an updated National Climate Change Policy, National Environment Policy, and National Agriculture Policy (SLYCAN Trust/APN 2019).

Further gaps and needs have been highlighted regarding the capacity to link NDC implementation in the country to the UNFCCC process and engage in thematic areas on adaptation and L&D. Furthermore, there is also a need to calculate losses and damages in the country and conduct a comprehensive assessment of finance and other means of implementation required to address them.

Like the other two countries, Sri Lankan stakeholders have also identified a need for institutional and human capacity building and enhanced coordination on NDC implementation, which would also require clear mandates, data-sharing policies, and access to information for relevant actors. Due to the institutional setup, there can be a perceived lack of ownership in some sectoral institutions, which hinders the building of synergies and efficient cross-cutting and cross-sectoral interventions connected to national development plans and priorities (SLYCAN Trust/ APN 2019).

Furthermore, there is a need for enhanced stakeholder engagement in climate change adaptation and L&D actions as part of NDC implementation and for amplifying the voices of vulnerable communities in decision-making processes, especially those with strong connections to ecosystems and natural resources. Access to finance and means of implementation are key constraints for NDC implementation, and contributions from the private sector were brought up as potential pathways for additional funding, particularly in the context of the current economic difficulties in the country.

MRV processes for climate actions in Sri Lanka remain in early stages. As in Bangladesh and Nepal, a need for the introduction of a budget-tagging system for adaptation and L&D finance was highlighted, as were the need for monitoring, evaluation, learning, progress tracking, and impact assessment systems (Mombauer 2019).

While the NDC sector on L&D contains commitments aimed at both risk management/reduction and adaptation, the NAP also includes explicit risk reduction components and refers to the L&D sector contained in the NDCs. Mainstreaming implementation of disaster risk reduction is envisioned through the sectoral climate adaptation cells, and the establishment of a coordination committee is proposed as a measure to enhance policy coherence and integration between disaster risk reduction and L&D.

3.4.4 Gender and Intersectional Considerations

Based on the initial NDCs, gender-responsive climate policies and plans have been marked as a need to be addressed, including awareness creation on gender-differentiated vulnerabilities, impacts, and implications of climate change; gender-disaggregated data collection and consideration of gender in planning and M&E processes; and incorporation of gender considerations into risk assessments and risk management frameworks. In addition, gender is only part of a complex set of intersectional vulnerabilities and must be considered in relation to marginalization, poverty, age, disability, ethnicity, and other cross-cutting identities and systemic factors (Wijenayake and Mombauer 2019).

While Sri Lanka's updated NDCs commit to "mainstreaming gender and social safeguards into adaptation priorities" and integrating "specific needs, vulnerabilities, and capacities of women, young children, disabled and elderly populations" through gender-responsive strategies and plans, there is still a need to strengthen this component in the implementation of the commitments. Key national plans, such as the National Adaptation Plan and the Climate Prosperity Plan, are gender-blind and do not correspond to the intersectional principles outlined in the NDCs.

For Bangladesh, the updated NDCs state that the planning process was conducted in a gender-responsive manner, although the document itself appears to be gender-blind. However, the country's new NAP contains a dedicated sub-section on implementation strategies related to gender, disability, youth, and social inclusion, presenting a key opportunity to incorporate these aspects into upcoming plans and actions. Similarly, Nepal's second NDCs commit to developing an action plan for gender equality and social inclusion that includes gender-specific programmes and resources, ensures gender-disaggregated data collection, and promotes participation, empowerment, and leadership of women, Indigenous people, and youth in climate change forums.

3.5 Common Gaps and Needs

Bangladesh, Nepal, and Sri Lanka share a range of commonalities when it comes to gaps and needs for NDC implementation of adaptation and L&D components. These shared gaps and needs can be clustered into the following six groups (SLYCAN Trust/APN 2019):

- Gaps in policies and laws, including the need for an overarching and comprehensive national legislation framework that facilitates implementation across all sectors and supports mainstreaming of climate change adaptation into sectoral processes;
- Gaps in institutions and coordination, including the need for a main coordinating entity with clear, proportionate, and adequate authority for the purposes of supervision over and provision of insightful leadership to other actors (such as sectoral institutions or local authorities);
- Gaps in capacities and awareness, including the need to enhance human and institutional capacities in key sectors and on all levels through well-designed tools, methods, and systems;
- Gaps in finance and technology, including the need to access sufficient and needs-based funding, technology, and other means of

implementation from different sources (such as multilateral funds, the private sector, and market-based mechanisms) while linking NDC implementation to national development plans;
- Gaps in data and information, including the need to enhance availability of high-resolution data, conduct assessments of capacity gaps, conduct assessments of loss and damage, establish knowledge-sharing platforms, and implement effective and transparent M&E and MRV systems;
- Socioeconomic and other factors, including the need to overcome socioeconomic challenges (such as poverty, youth unemployment, or population growth) and the need to incorporate and better address cross-cutting or intersectional vulnerabilities and issues (such as gender, age, disability, human mobility, or informal employment).

The need for more interconnected, coordinated, and inter-communicating systems and frameworks has emerged as a key commonality for all three countries, as well as a gap in terms of robust and transparent monitoring, evaluation, and learning systems that are fit for purpose, can guide implementation towards higher impact and adoption of best practices, and facilitate the continuous improvement of NDCs, NAPs, and national policy frameworks. Furthermore, all three countries have highlighted the importance of needs assessments, concrete and measurable targets, and progress tracking on adaptation (including budget tagging and tracking of climate finance) and L&D (SLYCAN Trust/APN 2019).

Finally, stakeholders stated the value of having publicly accessible systems and engaging multiple stakeholders as part of the NDC implementation process, including the private sector, civil society, community-based organizations, media, youth, the education system, and research institutions. However, there is currently a need to develop and establish more effective stakeholder engagement mechanisms and include other actors in planning processes, progress tracking, reporting, and support systems (SLYCAN Trust/APN 2019).

3.6 *Linking NDCs and NAPs to Other Relevant Processes*

As outlined in the gaps and needs analysis, there are key commonalities that focus on the establishment of systems for institutional coordination, knowledge-sharing, vertical and horizontal integration, M&E,

MRV, budget and progress tracking, L&D assessments, and stakeholder engagement.

As highlighted by many stakeholders, aligning climate change adaptation and L&D actions with other relevant processes can help to close these gaps and address the needs identified above. For example, the Paris Agreement, the SDGs, and the Sendai Framework were all signed in 2015 and have 2030 as a key target year. All three aim to reduce vulnerabilities and build resilience against a variety of risks and refer to multi-stakeholder-driven processes for achieving their targets. Therefore, implementation of adaptation and L&D components of NDCs could be effectively interlinked with SDGs and Sendai Framework to align common mandates, national targets, and data-sharing mechanisms and make effective use of finance and human resources (Handmer et al. 2019).

This includes the following:

- Linking NDC implementation and reporting on the subnational level with SDG localization processes and subnational implementation under the Sendai Framework (particularly priority area 2, 3, and 4) (SLYCAN Trust 2021);
- Aligning data collection and data-sharing mechanisms under the SDGs and the Sendai Framework with NDC review and implementation processes (SLYCAN Trust/APN 2019);
- Identifying co-benefits for accessing finance and support, including for capacity-building and awareness creation measures that could mainstream climate change adaptation, disaster risk reduction, L&D, and sustainable development at the same time and highlight interconnections and the continuum of interventions (ADB 2015);
- Aligning implementation of NDC adaptation components, NAPs, and other adaptation-related processes in the country to avoid duplication of efforts and enhance capacity-building, multi-stakeholder engagement, and sharing of knowledge and experiences among key stakeholders (SLYCAN Trust/APN 2019);
- Setting up a dedicated institutional framework for MRV to coordinate pre-existing data collection mechanisms and align them with reporting under the Paris Agreement, Agenda 2030, and the Sendai Framework, with a simplified and intuitive interface for data entry, local language data interface availability, in-built plausibility checks, and automated tracking of data in a transparent manner (Mombauer 2019);

- Identifying entry points to communicate support and capacity-building needs, lessons learned, good practices, and evidence from the NDC implementation process into relevant processes and forums on the regional (for example, SAARC) and global levels (for example, the Global Goal on Adaptation, the Global Stocktake, the Warsaw International Mechanism, the Santiago Network on Loss and Damage, the Glasgow Dialogue, funding arrangements for responding to L&D, the Paris Committee on Capacity-Building, the Nairobi Work Programme, Action for Climate Empowerment, Gender Action Plans, the Enhanced Transparency Framework, and others) (UNFCCC 2017).

3.6.1 Harnessing Synergies Between Adaptation, Disaster Risk Reduction, and L&D

Particularly in the context of NDC implementation and linking NDCs to NAPs, L&D processes, and actions under the Sendai Framework, stakeholders have highlighted the opportunity to identify and exploit synergies (UNDRR 2022). On one side, adaptation actions are closely connected to and often overlapping with disaster risk reduction and disaster management; on the other side, L&D is partially addressed through comprehensive disaster risk management (including risk transfer and risk retention instruments) (WIM Executive Committee 2016).

For example, Sri Lanka's NDCs explicitly place the development of a comprehensive risk management framework (L&D NDC 5) at the core of addressing L&D, while both Bangladesh and Nepal mainly address L&D through disaster management and in the context of residual risks. Conversely, elements of risk management are also included in the adaptation components of all three NDCs, for example, in seven out of nine of Sri Lanka's adaptation sectors (agriculture, livestock, water resources, coastal and marine, health, human settlements, tourism).

Building on these interconnections while acknowledging distinctions between different institutional setups, processes, and funding streams presents both a challenge and an opportunity for the future of planning and implementation for climate change adaptation and L&D.

4 Conclusion

NDCs are at the core of the Paris Agreement and represent critical cornerstones for both the national and global architecture of climate action. As developing countries with high vulnerability to climate change impacts and climate-induced L&D, Bangladesh, Nepal, and Sri Lanka have placed a significant focus on the adaptation and L&D components of their NDCs. Through the effective and successful implementation of commitments in these priority sectors, the three countries aim to enhance their adaptive capacities, mobilize funding for minimizing and addressing L&D, and develop coherent national frameworks for adaptation and climate risk management.

Developing countries such as Bangladesh, Nepal, and Sri Lanka face common barriers, constraints, obstacles, and needs for NDC implementation. By identifying these commonalities as well as potential solutions to address them, national action can be enhanced and accelerated across the South Asian region and beyond.

In particular, closing gaps related to policy coherence, inter-institutional coordination, and vertical and horizontal integration could enhance the effectiveness of implementation efforts. This includes not only actions and processes specific to adaptation and L&D but also the linkages and potential synergies between these processes, as identified in the existing documents. Climate change adaptation, disaster risk reduction, and L&D are closely related, and systems that harness co-benefits and complement each other have the potential to overcome many of the identified challenges, including those related to knowledge-sharing, stakeholder cooperation, and addressing cross-cutting and intersectional vulnerabilities.

Acknowledgements We wish to thank the researchers and team members of SLYCAN Trust, ICCCAD, and Prakriti Resources Centre as well as the national UNFCCC focal points, the relevant ministries, and their esteemed officers in Bangladesh, Nepal, and Sri Lanka for their support during the course of this research. Furthermore, we extend our gratitude to the officers of the adaptation and gender programmes of the UNFCCC Secretariat for their support in sharing valued knowledge and expertise to enrich the research. We also extend our appreciation to all consultation participants and other contributors to this research, including sectoral government officers in the three countries, technical experts, and resource persons.

An immense thank you to Dr. Linda Anne Stevenson, Ms. Nafesa Ismail, and Ms. Christmas Uchiyama for their guidance and the support provided in completing the activities of this workshop, despite difficult times in Sri Lanka. We are grateful for their cooperation and constant support in the activities of the research project.

References

ADB. (2015). *The Economics of Climate Change in South Asia: Adaptation and Impact Assessment*. Asian Development Bank.

Handmer, J., Stevance, A.-S., Rickards, L., & Nalau, J. (2019). *Achieving Risk Reduction Across Sendai, Paris, and the SDGs*. International Science Council.

IPCC. (2022). *Climate Change 2022: Impacts, Adaptation, and Vulnerability. Contribution of Working Group II to the Sixth Assessment Report of the Intergovernmental Panel on Climate Change*. Intergovernmental Panel on Climate Change.

Mombauer, D. (2019). *MRV Mechanisms Related to Nationally Determined Contributions and Reporting of Climate Action*. Colombo, Sri Lanka: SLYCAN Trust/APN.

SLYCAN Trust. (2021). *Research Brief: Integration of NDCs and SDGs for Climate*. Colombo, Sri Lanka: SLYCAN Trust (GTE) Ltd.

SLYCAN Trust. (2022). *Scoping Paper: Integrating Climate-Related Human Mobility into Nationally Determined Contributions and National Adaptation Plans. Human Mobility in the Context of Climate Change #5*. Colombo, Sri Lanka: SLYCAN Trust.

SLYCAN Trust/APN. (2019). *Policy Gaps and Needs Analysis for the Implementation of NDCs on Adaptation and Loss and Damage in Bangladesh, Nepal, and Sri Lanka*. Colombo, Sri Lanka: SLYCAN Trust.

UNDRR. (2015). *Addressing loss and damage at the national level: Lessons from Bangladesh* .

UNDRR. (2022). *Disaster Risk Reduction: A Vital Need in a Climate Changing World. SEM Policy Paper for COP 27.*

UNFCCC. (2012). *National Adaptation Plans. Technical guidelines for the national adaptation plan process*. LDC Expert Group.

UNFCCC. (2015). *Paris Agreement*. United Nations Framework Convention on Climate Change.

UNFCCC. (2017). *Report of the Executive Committee of the Warsaw International Mechanism for Loss and Damage associated with Climate Change Impacts. Addendum.*

UNFCCC. (2021). *National Adaptation Plans 2020. Progress in the Formulation and Implementation of NAPs.*

Wijenayake, V. (2019). *Integration of SDGs, the Sendai Framework, DRR, and NDCs for Effective Developmental Planning*. Colombo, Sri Lanka: SLYCAN Trust/APN. Retrieved from https://slycantrust.org/knowledge-resources/integration-of-sdgs-the-sendai-framework-drr-and-ndcs-for-effective-developmental-planning.

Wijenayake, V., & Mombauer, D. (2019). *Integrating Gender Responsiveness to Climate Change Adaptation Actions, and Moving towards Gender Responsive NDCs and NAPs*. Colombo, Sri Lanka: SLYCAN Trust/APN.

WIM Executive Committee. (2016). *Best Practices, Challenges and Lessons Learned from Existing Financial Instruments at All Levels That Address the Risk of Loss and Damage Associated with the Adverse Impacts of Climate Change*. Warsaw International Mechanism for Loss and Damage.

CHAPTER 13

Improving Climate Resilience of Urban Road Infrastructure in Southeast Asia

Lam Vu Thanh Noi, Richard T. Cooper, Tran Quang Minh, Cao Thi Thu Huong, Spoann Vin, Sath Sitak, Rotchana Intharathirat, Jaranporn Lertsahakul, and Tran Thi Tinh

Abstract Road infrastructure plays an important role in overall national socio-economic development. Climate change poses an increasing risk to urban roads, and municipal authorities lack the know-how in assessing loss and damage that may arise from flood events and in the selection of

L. V. T. Noi (✉)
Southern Institute of Water Resources Research, Ho Chi Minh City, Vietnam
e-mail: lamuem1980@gmail.com

R. T. Cooper
Southeast Asia START Regional Centre, Chulalongkorn University, Bangkok, Thailand

T. Q. Minh
Sub-Institute of Hydrology, Meteorology and Climate Change, Ha Noi Capital, Vietnam

V. Wijenayake et al. (eds.), *Linking Climate Change Adaptation, Disaster Risk Reduction, and Loss & Damage*,
https://doi.org/10.1007/978-981-99-8055-0_14

relevant adaptation options. This paper determines climate change adaptation measures and strategies for improving climate resilience of urban road infrastructure in Southeast Asia. A spatial assessment methodology using NK-GIAS software is designed to assess loss and damage of urban roads related to flooding in three cities of Cambodia, Thailand and Vietnam. Multi-Criteria Analysis was used to assess adaptation measures. Economic costs related to different flood scenarios were estimated for each of the selected cities. Road networks were found to be at particular risk from flood events. Under the most extreme flood scenario, maximum road damage was estimated at 3 million USD for Kampot, 20 million USD for Hoi An and 21 million USD for Samut Sakhon. It is proposed that the spatial analyses could be improved in future work through the development of damage curves to address both the duration and depth of flooding.

Keywords Road · Climate change · Climate resilience · Risk assessment · Climate change adaptation

---

C. T. T. Huong
Transport Development and Strategy Institute, Hanoi, Vietnam

S. Vin · S. Sitak
Department of Economic Development, Faculty of Development Studies, Royal University of Phnom Penh-RUPP, Phnom Penh, Cambodia

R. Intharathirat
Environmental and Pollution Control, Pollution Control Department, Ministry of Natural Resources and Environment, Regional Environmental Office 9, Office 9, Udon Thani, Thailand

J. Lertsahakul
Thailand Green Style Co., Ltd., Bangkok, Thailand

T. T. Tinh
Dalat University, Da Lat, Lam Dong Province, Vietnam

Highlights

- This study presented the status of road infrastructure in the context of climate change in six cities of Vietnam, Cambodia and Thailand.
- Results from Rapid Vulnerability Assessment in the six cities indicate that the main climate change-related factors affecting roads are flood and SLR.
- This study demonstrates how economic losses can relate to different flood scenarios in the three selected cities.
- The development of road damage curves by laboratory analysis is recommended for future research.

1 Introduction

Climate change impacts across all sectors and groups, requiring the implementation of appropriate adaptation measures for any given type of impact, sectors and groups, which pose challenging conditions in many countries, but especially developing ones. Historical evidence and future climate change projections point to past and ongoing climate change on a global scale, especially in Southeast Asia. Increases in average annual temperature, changes in regional and local precipitation regimes and sea level rise (SLR) in the coastal regions were evident during the last century. Future climate projections indicate increasing temperatures, precipitation and sea level rise of up to 40 cm by 2100 (IPCC 2013, 2014; Noi and Nitivattananon 2015). Increased unpredicted frequencies of weather extremes, tsunamis, tropical cyclones, floods, saltwater intrusion and drought have been experienced in Southeast Asia (ADB 2009, 2010; Noi and Nitivattananon 2015). Scientific projections indicate that climate change will lead to increased precipitation extremes as well as flooding and coastal erosion caused by SLR will become a major hazard impact on coastal communities (Un Habitat 2011; Intergovernmental Panel on Climate 2014; Noi and Nitivattananon 2015).

Climate change impacts all sectors across Asia, and communities will be increasingly vulnerable to the effects of climate change through its impact on infrastructure, agricultural production, water and other natural resources (IPCC 2014). Given the region's high population, rapid urbanisation and development, and economies reliant on agriculture and the use of natural resources, Southeast Asia is recognised as a region that is vulnerable to the effects of climate change (ADB 2009, 2010). People are increasingly affected by flooding in low-lying areas of Southeast Asia. Potentially up to five million people are at risk in the Mekong River delta alone (IPCC 2007, 2014) due to SLR, with groundwater abstraction in the megacities of Bangkok, Ho Chi Minh City (HCMC) further contributing to the risk of flooding (Hens et al. 2018). Appropriate adaption options need to be selected to address the vulnerability of infrastructure to SLR (Hens et al. 2018) and other hazards such as flooding, drought, landslides and erosion (World Bank 2010).

Hard and soft adaptation options offer potential solutions to address climate risks. Hard measures include strengthening existing and building new infrastructure, including the adoption of revised design standards for roads and their appurtenances (ADB 2010). Soft measures might include improving early warning systems, emergency response, policy and planning (World Bank 2010). Urban roads, as well as other infrastructure, are considered priority areas for investment in Southeast Asia to address climate change impacts (ADB 2009).

This chapter focuses on the vulnerability of urban roads to flood in the context of climate change. It adopts the definition of climate risk as the interaction between hazard, vulnerability and exposure per the IPCC Sixth Assessment Report (Reisinger et al. 2020). IPCC defines flood risk in the context of the flood hazard (the magnitude and/or frequency of flooding), the exposure of the system affected (e.g. infrastructure and other systems in the area potentially affected by flooding) and the vulnerability of the system (e.g. design, maintenance and operation of infrastructure, multi-hazard early warning systems) (Reisinger et al. 2020). However, the adaptive capacity to respond and recover from the impacts of climate change is important to consider. The International Strategy for Disaster Reduction (ISDR 2002) defined vulnerability as the opposite of capacity, so that increasing capacity means reducing vulnerability, and high vulnerability means low capacity. In this context, the term "vulnerability" refers to the vulnerability of people, areas, institutions and sectors within the study area with respect to climate change.

Vulnerability can be separated from the concept of adaptive capacity, enabling the components of risk (Reisenger et al. 2020). However, ISDR (2002) and Noi and Nitivattananon (2015) define risk as expressed by the equation Risk = Hazards x Vulnerability/Capacity. It does not conflict with the IPCC definition of vulnerability because vulnerability and adaptive capacity are still closely related, as good adaptive capacity will reduce vulnerability and risk in the study area. This paper is based on the ISDR concept of risk for assessing loss and damage of urban roads in relation to various flooding scenarios (different flood depths) in three cities of SEA.

Urban planners need an appreciation of the vulnerabilities of urban transportation infrastructure (UTI) to climate change. The selection of practicable adaptation options should reflect current best practices in addressing climate resilience, with decisions not only based on cost. Vulnerability assessments can improve decision-making by highlighting potential hazards, and vulnerabilities and the selection of appropriate adaptation measures to minimise loss and damage. To determine the vulnerabilities and sensitivities of UTI in relation to various climate change scenarios, this research applies various vulnerability assessment techniques and tools (Füssel and Klein 2006; Noi and Nitivattananon 2015).

This study aims to improve climate change adaptation for UTI planning in selected cities of Southeast Asia through contributing new knowledge and capacity building. The chapter aims to advance knowledge on loss and damage of roads in the context of climate change and determine climate change risk assessment approaches. There are two key aims of the study: (i) to develop a method for determining loss and damage to urban roads in the context of climate change and at the subnational level; and (ii) to determine climate change adaptation measures for enhancing climate resilience of urban road infrastructure pertinent to coastal cities of Southeast Asia.

As per the ISDR concept of risk, this paper examines the linkages between hazards, vulnerabilities and adaptive capacities, focusing on flood-related loss and damage for urban roads, and the identification of appropriate adaptation options for urban transport planning at the city level. There were two key limitations in the study: firstly, only negative impacts of climate change were considered, and secondly, given that no laboratory studies were undertaken, the study used existing road damage curves derived from the literature review.

2 Methodology

Six cities, including Hoi An and Vinh Long in Vietnam, Huahin and Samut Sakhon in Thailand and Sihanoukville and Kampot in Cambodia, were selected for conducting rapid assessments of loss and damage of transportation infrastructure. Three techniques were used to assess loss and damage in the context of climate change: Multiple Criteria Analysis (MCA), Impact Matrix and Participatory Rural Appraisal (PRA). NK-GIAS software was used to conduct vulnerability analyses of three cities in Thailand, Cambodia and Vietnam following a rapid assessment process. Loss and damage was estimated for each type of UTI related to climate change-related hazards. The overall methodological framework is presented in Fig. 1.

Figure 1 shows the four phases of the NK-GIAS (Samarakon L. et al. 2012) road damage estimating procedure. The research defined climate change-related disaster risk as a function of the interdependence of hazards, vulnerability and adaptive capacity ISDR (2002)

$$\text{Risk(R)} = \text{Hazard(H)} \times \frac{\text{Vulnerability(V)}}{\text{Capacity(C)}} \tag{1}$$

In step 1a, six cities in Vietnam, Thailand and Cambodia (two per country) were selected for conducting a rapid vulnerability assessment for urban roads in the context of climate change. City selection criteria were based on climate change impacts on infrastructure and the availability of secondary data on climate change-related hazards and loss and damage of urban roads. **For step 1b,** the urban roads in each city were classified into the main road types based on field survey, literature review and consultation with key stakeholders.

Step 2 involved conducting a Rapid Vulnerability Assessment (RVA) to determine recent loss and damage to urban roads using secondary data over the previous 5–10 years. To identify vulnerabilities and adaptive capacities, the assessments employed PRA, Impact Matrix and Multi-Criteria Analysis (MCA).

Based on the results from the previous tasks, in **step 3**, the most vulnerable city in each country was selected as the pilot city for detailed vulnerability assessments to estimate loss and damage for the main urban road types. Road damage was estimated using NK-GIAS software (Samarakon L. et al. 2012). The damage calculation process utilising NK-GIAS is outlined as follows: (1) processing of topographic, meteorological and hydrological baseline data; (2) creating road map density layers; (3) creating hazard profiles with different flooding scenarios; (4) simulating hazard maps with future projected maximum flood depths; (5) based on secondary data, survey results and literature reviews, building on the damage curves of city roads; (6) using the function of total structural

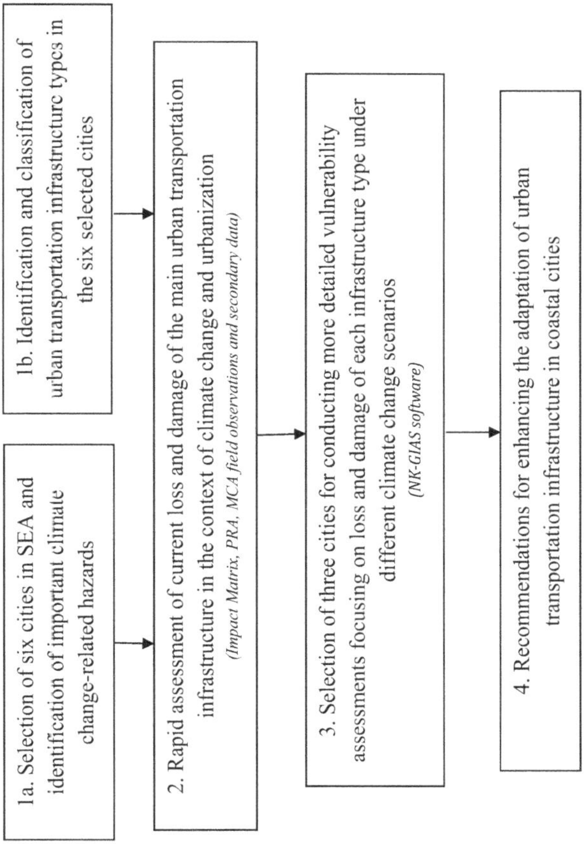

Fig. 1 Overall methodological framework

damage for each type of road based on the results of steps (3) and (4), calculating damage for each type of road; and (7) summarising damage cost estimates.

In step 4, through convening various meetings, research findings were shared with key stakeholders including municipalities, provincial transportation departments and urban planning departments, and recommendations were proposed for enhancing climate change resilience of urban roads in coastal cities.

3 Results and Discussion

3.1 Rapid Vulnerability Assessment of Urban Roads in Selected Cities

Vietnam, Thailand and Cambodia were selected as the study areas in Southeast Asia, based on consideration of their vulnerability to climate change, especially to SLR and flooding, and the authors' research networks in the region. Additionally, representation, the magnitude of climate change impacts and data accessibility were criteria used for choosing the cities in this research. Based on the literature review results, the study areas selected were Hoi An and Vinh Long (Vietnam), Hua Hin and Samut Sakhon (Thailand) and Kampot and Sihanoukville (Cambodia). The results of the literature review and field surveys in the three cities are presented in Table 1.

As approximately 70% of Hoi An is located below 3 m elevation, it is extremely vulnerable to SLR (Hoi An PC 2015; SIHYMECC 2015). It is acknowledged that climate change adaptation measures are required to reduce threats to natural resources, the local community, the economy and transportation assets. Currently, the city is concentrating on adaption strategies to lessen coastal erosion.

Increasing sea levels and rainfall have the potential to worsen flooding in Vinh Long (2015). Given its flat geography, the city's UTI is more susceptible to floods, and hard and soft adaptation strategies have so far proved to be inadequate.

A popular tourist spot on the western edge of the Gulf of Thailand, Hua Hin, suffers from pluvial flooding that will likely worsen in the future due to heavier monsoonal rainfall. Furthermore, as the Thailand Meteorological Department (2012) reported, the region is at "relatively high risk" from typhoons and tropical storms. Current adaptation options do not

Table 1 Key features of the six selected cities

City	*General information*	*Urban transportation infrastructure*
Hoi An, Vietnam	– Total area: 6,148 km^2 (Hoi An PC 2022) – Total population: 98,599 (Hoi An PC 2022) – Located downstream of the Vu Gia-Thu Bon River, a sensitive coastal area – Designated as a World Heritage Site since 1999. The city is especially vulnerable to climate change, with 70% of the city lying less than 3 meters above sea level (SIHYMECC 2015)	– The main inland road system is 5–7 m wide and in some places up to 14–20 m wide – The inner city road system is narrow and old, as is the supporting infrastructure. The total length of the road is 133.6 km, with 67% being bituminous (Hoi An Transportation Department 2017) – Based on the road surface structure classification, the main roads include bituminous, penetration macadam, concrete, asphalt treatment, and earth (Hoi An Transportation Department 2017)
Vinh Long, Vietnam	– Total area: 47.82 km^2 (Vinh Long PC 2022) – Total population: 1,022,791 (Vinh Long PC 2022) – Located between the Tien and Hau rivers in the Mekong Delta's central region – At risk to rising sea levels and heavy rainfall (Vinh Long PC 2022)	– The city has access to water and inland road networks given its location in an area with a dense stream network – Flash flooding with depths of up to 20–30 cm cause traffic congestion and damage. These roads fail to meet the required rain drainage standard (Vinh Long PC 2022) – Because of the city's lowland location, it is vulnerable to flooding of increasing duration and magnitude, with UTI already affected (Vinh Long PC 2022) – Hard and soft climate change adaptation measures are being implemented, but those for UTI are insufficient (Vinh Long PC 2022)

(continued)

Table 1 (continued)

City	*General information*	*Urban transportation infrastructure*
Hua Hin city, Thailand	– Total area: 86.36 km^2 – Total population: 85,099 (Hua Hin Municipality 2015) – Popular coastal resort and the largest coastal settlement in Prachuab Khiri Khan province – Hua Hin receives a large number of tourists and has a rapidly growing population (Hua Hin Municipality 2015) – With the increasing rainfall extremes, inland flooding caused by runoff is likely to worsen in the future. Flash flooding is a problem in low-lying areas during the rainy season (Hua Hin Municipality 2015; JICA 2013)	– Road and rail comprise the key UTI in Hua Hin – Climate change adaptation efforts are underway, but they are not directed at UTI (Hua Hin Municipality 2015; JICA 2013) – Transportation planning, according to the Hua Hin Municipality (2015), is limited to a road 'densification' target of 1 km per km^2 (currently 0.8 km/km^2)
Samut Sakhon city, Thailand	– Total area: 1,033 km^2 – Total population: 68,380 (Samut Sakhon Municipality 2018) – Meuang Samut Sakhon, the capital district of Samut Sakhon province, is part of the Bangkok Metropolitan Region and is particularly important to this research. The province is divided into three districts – Coastal flooding caused by sea level rise, storm surge, and coastal erosion (Samut Sakhon Municipality 2018)	– Key UTI include road and rail networks, as well as related assets – Climate adaptation measures are being implemented, focusing on hard infrastructure solutions (Green Style Co. 2018; Samut Sakhon Municipality 2018) – Construction of sea walls and ground elevation are engineering adaptation solutions, albeit the latter has been known to leak (Samut Sakhon Municipality 2018)

City	*General information*	*Urban transportation infrastructure*
Sihanoukville city	– Total area: 85.56 km^2 – Total population: 156,691 (RUPP 2018) – The city of Sihanoukville is a port and is known for its tourism – Sihanoukville is described as an economic zone port and tourism development city – In a previous study of Sihanoukville's vulnerability (Un Habitat 2011), the rise in sea level could reach 1.5 m by the end of the century and poses a threat to low-lying areas (JICA 2015; MPWT 2013)	– Four districts in Sihanoukville have a total road length of 48,385 km, of which 3,668 km are made of concrete, 25 km from laterite, 4.8 km from created earth, and 4.74 km from unconstructed earth (RUPP 2018) – As reported by MPWT (2013), the railway line that runs between Phnom Penh and Sihanoukville is 266 km long – The local government spends more than USD 50,000 annually on road repairs and maintenance (MPWT 2013)
Kampot	– Total area: 39.41 km^2 – Total population: 35,874 (RUPP 2018) – This city is susceptible to a number of climatic risks, such as strong winds, rising sea levels, extended drought, saltwater intrusion, and pluvial flooding. Historical data illustrates the damage that such extremes can do to city infrastructure and its inhabitants (JICA 2015; MPWT 2013) – Several communes, including Chum Kriel, Traeuy Koh, Boeung Sala Tboung, and Russei Srok Keut, are extremely vulnerable to sea level rise. Flooding of salt farms and saltwater intrusion are two major consequences of sea level rise (MPWT 2013)	– Kampot's road system is 45,117 km long and spans five districts, with 1,556 km of concrete, 33.224 km of laterite, 2.848 km of earth, and 4.760 km of unconstructed earth (RUPP 2018) – Roads, dikes, canals, bridges, railways, and drainage and water supply systems are all vulnerable to storms, floods, and sea level rise. These hazards cause significant economic, social, and infrastructure damage (RUPP 2018)

address UTI, even though climate hazards have the potential to increase maintenance costs.

Situated on the Gulf of Thailand's northern shore, Samut Sakhon is susceptible to a range of climate-related hazards including SLR, saltwater intrusion, storm surge and coastal erosion. Hard infrastructure improvements are being prioritised as part of the implementation of adaptation strategies (Samut Sakhon Municipality 2018).

Kampot is located on the Gulf of Thailand in the southwest of Cambodia. Storms, extreme rainfall, floods and saltwater intrusion are all threats to Kampot (JICA 2015; MPWT 2013). These hazards directly impact on vital services such as water supply, sanitation and road infrastructure, all of which are underdeveloped. Climate change has previously been studied in terms of its effects on urban infrastructure: roads, dikes, canals, bridges and railways, drainage systems and water supplies are all vulnerable to storms, floods and SLR (JICA 2015; RUPP 2018).

Sihanoukville City is a southern coastal city of Cambodia and the capital of Preah Sihanouk Province. Sea level rise, storms and flooding contribute to infrastructure losses including UTI damage (JICA 2015; MPWT 2013), and current coastal infrastructure is at increased risk from climate-related impacts. It is challenging to lessen the effects of flood disasters due to the lack of knowledge and money, as well as the inadequate quality (such as earthen roadways) and quantity of UTI. The city's vulnerability has been rated as "medium", but some areas along the coastline are considered extremely vulnerable.

Each city underwent a rapid vulnerability assessment. The research team gathered information from interviews with local experts and detailed the exposure to climate change-related risks, UTI vulnerabilities and each city's characteristic geographic, social and economic features. Current and potential adaptation and adaptive capacity approaches were discussed, and each city's vulnerability was ranked accordingly. A city was assessed as having the maximum vulnerability if no adaptation measures were in place and as having "low vulnerability" if some adaptation measures were being implemented. If a city was determined as having a strong capacity for adaptation, it was said to have "low vulnerability" or "negligible vulnerability" if there were no climate-related dangers.

A summary of RVA results is given in Table 2.

Based on an RVA that included group discussions and expert interviews, Table 2 shows the sensitivity of the six chosen cities and their capacity to adapt to existing and future climate-related impacts. The table

Table 2 Summary of RVA results for the selected cities

Country/ City	*Vulnerability*	*Hazards*							
		Storm	*Flood*	*SLR*	*Saline intrusion*	*Drought*	*Coastal Erosion*	*Land Subsidence*	*Land slide*
Hoi An, Vietnam	Level of road vulnerable	+++	+++	+++	-	+	+++	+	-
Vinh Long,Vietnam	Level of road vulnerable	+	+++	++	+	+	+++	+	++
Kampot, Cambodia	Level of road vulnerable	++	+++	++	-	+	++	–	+
Sihanoukville, Cambodia	Level of road vulnerable	++	++	++	+	+	++	-	+
Hua Hin, Thailand	Level of road vulnerable	+	+	++	+	+	++	+	+
Samut Sakhon Meuang, Thailand	Level of road vulnerable	+	+++	+++	-	+	+++	+	-

Note: (-) **Negligible** + **Low** ++ **High** +++

Source Experts' opinions and survey outputs (2015)

shows that when a city's adaptive capacity is low, the city is vulnerable or highly vulnerable to hazards. In contrast, when the adaptive capacity is medium or higher, the city may be vulnerable but not highly vulnerable to hazards. Vulnerability increases when adaptive measures are lacking. All hazards were evaluated in terms of their impacts on the urban roads. The effectiveness of implementing a climate change adaptation measure for the Samut Sakhon River was demonstrated using evidence derived from field observations and expert interviews.

Table 2 shows that SLR and floods are the primary climate change-related concerns affecting urban roads, and they similarly present potential risks for roads in the other chosen cities. Based on RVA results, the three cities (one in each country) identified as having the highest road vulnerability—Hoi An, Kampot and Samut Sakhon—were chosen for a more detailed vulnerability assessment focused on road loss and damage under various flood scenarios.

3.2 Assessment of Road Damage by Applying NK-GIAS

3.2.1 Damage Curve Development for Road Damage

According to De Bruijn (2005), the damage function for infrastructure is defined as follows:

$$ED_{infrastructure} = \alpha(d).D_{max}.\varepsilon \quad (2)$$

where

ED is economic damage in US dollars (USD)
α presents the damage factor
d indicates water depth (meter)
D_{max} presents the maximum damage per unit in the category (USD/m^2)
ε shows the conversion/extra factor

The damage function was adapted in this paper to calculate road damage as follows:

$$\text{Road D} = \sum(\alpha_i.D_i.S.\varepsilon) \quad (3)$$

where

- $D_{i\ max}$ is the maximum damage per m^2 (construction cost, USD/m^2)
- S presents cell size (m^2)
- α_i shows the roads damage factor, it is flood depth (h).
- $\varepsilon = 1$: Roads were digitised using detailed road and elevation maps. Because the roadways can be measured precisely, there is no need for a correction factor and $\varepsilon = 1$.

The function shows the relationship between flood depth and road damage, as provided by De Bruijn (2005) and Jorik Chen (2007), and adapted for calculating the road damage factor (α_i). The α_i and flood depth (h) are correlated as per the road damage curve (4):

$$\text{If } x(h) < 0.91\text{ m}, y(\alpha) = 5.581x^4 - 7.9492x^3 + 4.4176x^2 - 0.5439x + 0.0018 \quad (4)$$

$$\text{If } x(h) \geq 0.91\text{ m}, y(\alpha) = 1$$

Table 3 Road damage factors (α_i)

x (Flood depth in metres)	*Υ* (α)
0	0.00
0.15	0.00
0.35	0.10
0.60	0.27
0.75	0.49
0.91	1.00

Sources Adapted from De Bruijn (2005) and JorikChen (2007)

The α_i values were calculated using flood depth, and the results are shown in Table 3.

The α_i value was used to calculate road damage and is based on flood depth in the cost damage calculation.

3.2.2 *Results of Maximum Road Damage Under Different Flood Scenarios*

The results of road damage based on flood scenarios are summarised in the following sections. The methodology mentioned in Sect. 2 was fully applied for calculating road damage using NK-GIAS. Using historical flood depth data and the actual road damage in the case study sites, the NK-GIAS programme was validated and calibrated. The model was validated and calibrated before being applied to the three chosen cities, with real road damage compared to calculated values of maximum road damage.

3.3 a. Flood Scenarios

The highest historical water levels were derived from data available from the previous 10 to 20 years and used to inform the flood scenarios.

Flood scenario analyses used the following relationship:

$$\textbf{Water level } (\mathbf{H_{max}}) = \textbf{flooding depth (h, meter)} + \textbf{ground elevation (meter)} \quad (4)$$

Historical H_{max} values included the following:

- $H_{max} = 3.40$ m in Hoi An on November 1964 (SIHYMECC 2015);

- H_{max} = 4.0 m in Kampot on September 2013 (JICA 2015); and
- H_{max} = 2.7 m in Samut Sakhon on October 2011 (Green Style Co. 2018; Samut Sakhon Municipality 2018).

As noted above, the maximum road damage was estimated for new flood scenarios by increasing Hmax and basing these on the maximum flood levels of the last 10 to 20 years.

The proposed scenarios for estimating the road damage are as follows:

Scenario 1 (SCE 1): The average water level in the study locations for the previous 20 years.

Scenario 2 (SCE_2): The water level equal to the maximum flood depth experienced in the study areas over the last 20 years.

Scenario 3 (SCE_3): The water level that corresponds to a 3 to 7% increase in flood depth over the last 20 years in the study locations (based on the topography of each city and local expert consultation).

Scenario 4 (SCE_4): The water level in the pilot study increased by 10 to 17.5% when compared to the maximum flood depth over the last 20 years (based on local expert consultations and recording data).

4 b) Maximum Road Damage Calculation.

The cost of restoring the road to its pre-flood condition is considered road damage.

- **Hoi An City**

Table 4 details the main urban roads investigated in the study. The main roads are divided into asphalt concrete, penetration; cement concrete and surface dressing pavement with total road length, surface, construction cost and the percentage of each kind of road in Hoi An are presented as follows (Figs. 2, 3, and 4).

Figure 3 depicts road damage factors (α_i) determined in Hoi An under scenario 1 (SCE 1), and Table 5 depicts road damage costs for each flood scenario and type of urban road.

According to Hoi An transportation department (2017), an analysis of secondary data and expert consultation revealed that damages mostly relate to the road surface. As discussed during a consultation with the transportation expert, the road foundation can reduce flood damage

Table 4 Road types in Hoi An City

Roads	*R1 (AC, SDP, CC)*	*R2 (Pen, AC, CC, Earth)*	*R3 (AC, Pen, CC)*	*R4 (AC, Pen, CC)*	*R5 (AC,CC, Pen)*	*Total*
Length (m)	84,024	20,742	7,078	6,146	15,620	133,610
Percentage of length (%)	62.89	15.52	5.30	4.60	11.69	100.00
Surface (m^2)	1,247,859	239,209	62,936	45,201	71,195	1,666,401
Percentage of surface (%)	74.88	14.35	3.78	2.71	4.27	100.00
Cost (USD/m^2)	90	90	82	68	38.83	

Note AC: Asphalt concrete; CC: Cement concrete; Pen: Penetration; and SDP: Surface dressing pavement; VMC: Vietnam Ministry of Construction
Source Hoi An Transportation Department (2017) and Vietnam Ministry of Construction (2015)

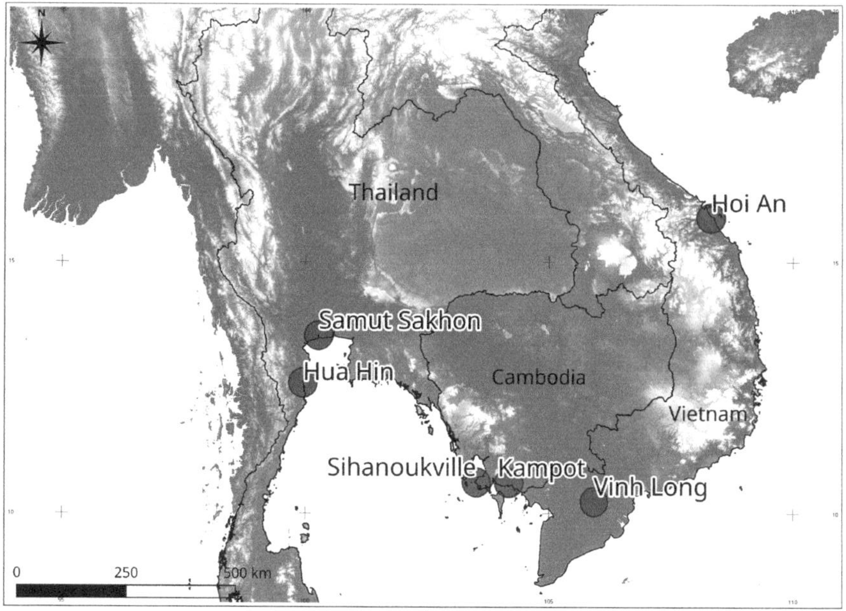

Fig. 2 Locations of six cities where research was conducted

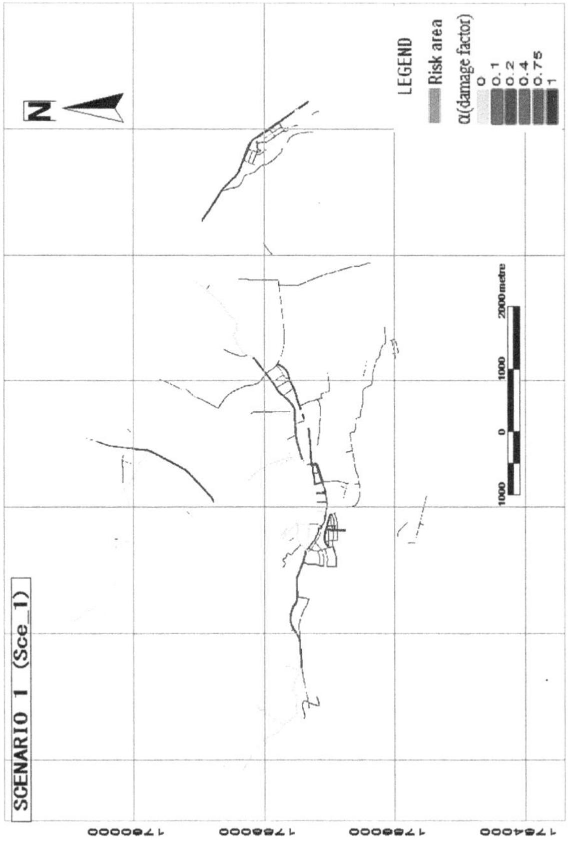

Fig. 3 Flood risk map for urban roads under Scenario 1 in Hoi An City

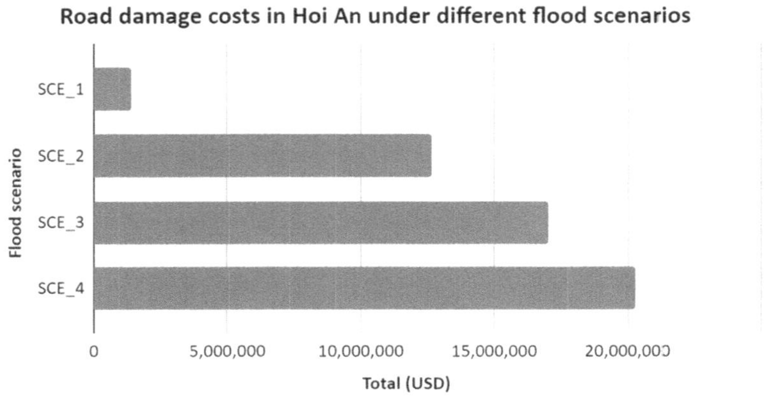

Fig. 4 Total road damage cost under different flood scenarios in Hoi An City

Table 5 Total road damage in Hoi An City under various scenarios

Name of scenario	*SCE_1 (m)*	*SCE_2 (m)*	*SCE_3 (m)*	*SCE_4 (m)*
Road damage	**2.72**	**3.4**	**3.61**	**3.94**
R1	621,335	6,173,497	8,886,456	11,344,227
R2	74,411	776,228	1,165i,390	1,528,950
R3	71,559	852,295	1,187,013	1,380,114
R4	550,092	3,992,492	4,562,281	4,631,773
R5	104,912	884,930	1,220,152	1,386,971
Total (USD)	1,422,311	12,679,444	17,021,294	20,272,036

through robust road construction and other adaptation measures such as dikes and embankments. This study does not apply new hydraulic modelling and only integrates the hydraulic effects based on the road damage curve. According to the findings, the maximum road damage under flooding scenario 4 is 20.3 million USD (Table 5).

- **Kampot city**

The characteristics of Kampot's roadways are shown in Table 6.

Table 6 Main types of roads in Kampot City

Road types in Kampot City					
	R1 (AC)	*R2 (CC)*	*R3 (DBST)*	*R4 (Laterite)*	*Total*
Length (m)	**9,295**	3542	14,150	21,945	48,932
Surface (m^2)	103,182	37,246	142,017	149,952	432,397
Percentage of Length (%)	19%	7%	29%	45%	
Percentage of Surface (%)	24%	9%	33%	35%	
Cost (USD/m^2)	48	44.18	34.4		

Note The DBST: Double Bituminous Surface Treatment
Source Kampot Administration (2015)

Figure 5 depicts road damage factors for SCE_1, and Fig. 6 and Table 7 depict damage costs for each type of road in Kampot under the flood scenario.

Each scenario's overall cost of loss and damage from urban floods is considerable (Table 7). The lowest economic loss of the three scenarios was experienced by SCE_1, with expenses totalling more than 952 K USD with an average flood depth of 0.3 m. The cost of road damage for SCE 2, which simulates the worst flood in the last two decades, is 2.7 million USD more than the cost of the base scenario and similar to the damages for scenario 3. However, the economic loss climbs to over 3 million USD when the water level rises by 17.5% with SCE_3.

The findings show how different road surfaces can affect costs. When compared to other road types, asphalt concrete is the most vulnerable to flood. Due to limited resources and geographical availability, cement concrete roads are the preferred choice for commune road development. Laterite roads were mostly used in peri-urban communities to connect major city roads. Although laterite roads account for 45% of the total road length in the city, their economic loss is lower when compared to other types. In the city, cement concrete roads are more susceptible to damage from flooding than DBST roads, with the latter being the most typical road type.

Figure 6 presents the cost of road damage under various scenarios. The results show that the base scenario is less vulnerable under the current road infrastructure development situation. Alternative scenarios significantly affect UTI, particularly for AC and DBST road types. Due to the

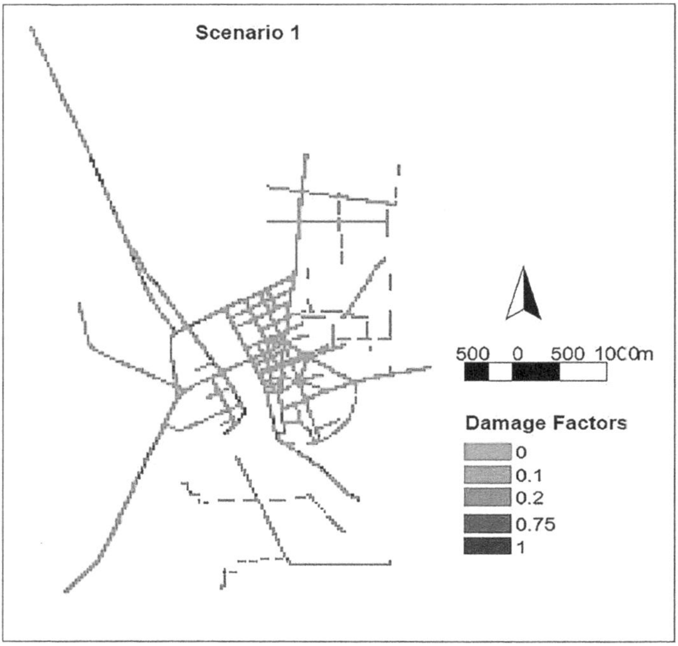

Fig. 5 Flood risk map for urban roads under Scenario 1 in Kampot City

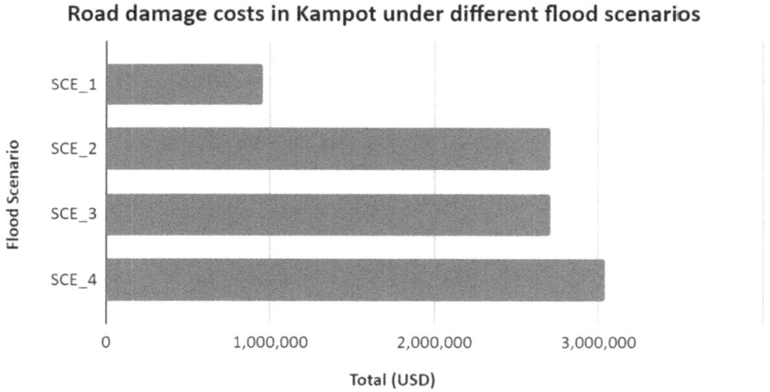

Fig. 6 Total road damage costs under different flood scenarios in Kampot City

Table 7 Total road damage in Kampot City under various scenarios

Scenario	*SCE_1 (m)*	*SCE_2 (m)*	*SCE_3 (m)*	*SCE_4 (m)*
Road damage	**3.3**	**4.0**	**4.3**	**4.7**
R1	516,987	1,623,104	1,623,104	1,647,624
R2	27,387	223,745	223,745	508,044
R3	393,855	758,393	758,393	779,599
R4	13,860	104,445	104,879	105,940
Total cost (USD)	952,089	2,709,688	2,710,122	3,041,208

limitations of the DEM used in Kampot city, estimated damage costs did not differ significantly among scenarios 2, 3 and 4. (30 m resolution). A higher DEM resolution would have most likely improved results.

- **Samut Sakhon City**

Table 8 shows the various metrics for road infrastructure in Samut Sakhon.

There are only three types of urban roads in Samut Sakhon: asphalt (A), reinforced concrete (RC) and highway with RC surfaces. The total length of the highway road is 39,215 km, accounting for approximately 9% of the city's road surface.

Figure 7 depicts the road damage factors for Scenario 1, and Table 9 and Fig. 8 depict the estimation of road damage under various flood scenarios.

Table 8 Main kinds of road in Samut Sakhon City

Road types in Samut Sakhon	*R1 (RC)*	*R2 (A)*	*R3 (Highway)*	*Total (m)*
Length (m)	**18,233**	23,029	39,215	80,477
Surface (m2)	332.20	272.80	63.00	668
Length percentage (%)	23%	29%	49%	
Surface percentage (%)	50%	41%	9%	
Cost (USD/m^2)	72,5	29	300	

Note RC: Reinforced Concrete
Source Samut Sakhon Municipality (2018)

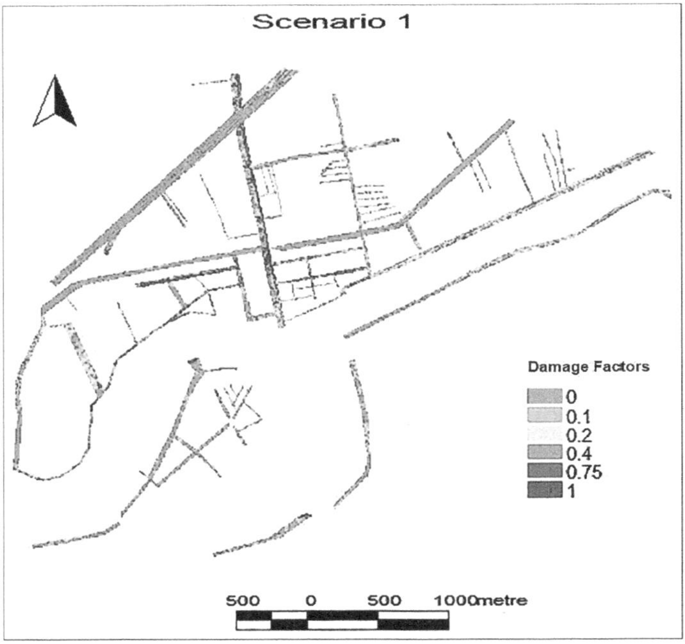

Fig. 7 Flood risk map for urban roads under Scenario 1 in Samut Sakhon City

Table 9 Total road damage in Samut Sakhon under various scenarios

Scenario	*SCE_1 (m)*	*SCE_2 (m)*	*SCE_3 (m)*	*SCE_4 (m)*
Road damage	**2.4**	**2.7**	**2.79**	**2.99**
R1	1,921,350	3,912,921	4,409,701	5,272,966
R2	3,534,049	6,934,664	7,776,366	9,069,439
R3	1,139,290	3,698,020	4,528,451	6,703,217
Total cost (USD)	6,594,689	14,545,605	16,714,518	21,045,622

Samut Sakhon is a low-lying city with a mean elevation of less than 1.8 m (Samut Sakhon Municipality 2018). Road damage costs were based on SCE_3 with a water level of 2.79 m, representing a level 3.3% greater than the historical maximum (H_{max}) in October 2011 (Green Style Co. 2018). The city will be protected by continuing the construction of a

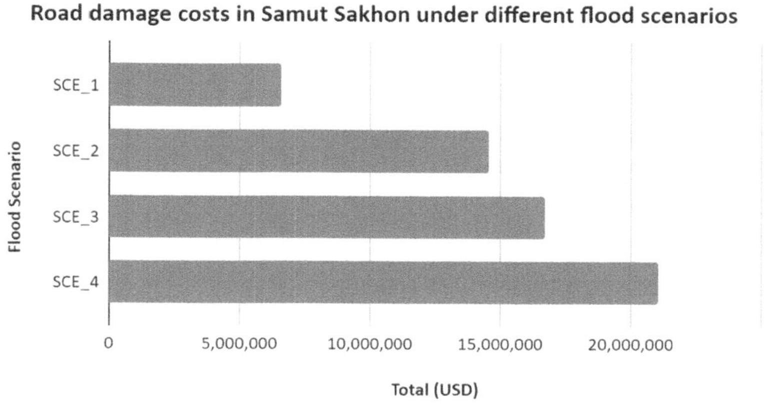

Fig. 8 Total road damage costs under different flood scenarios in Samut Sakhon City

river dike, which is a crucial climate change adaptation measure in the study region. However, with just 0.09 m greater flood depth in SCE_3, the economic damage increases by 15%, from 14.5 to 16.7 million USD. In SCE_4 compared to SCE_1, the flood depth rises by 0.29 m.

The results show that the current state of road infrastructure development is vulnerable under SCE_1 From conversations with the Disaster Prevention Department (2018), Tha Shalom and Mahachai subdistricts were inundated by a mean flood level of 0.6 m (water level of 2.4 m) before the dike was built. Flood depths are currently less than 0.6 m because of dike construction that took place between 2007 and 2009, with the exception of the Chalom subdistrict, which lies beyond the dike.

Alternative flood scenarios have greater effects on urban roads, particularly for RC roads. Due to differences in inundation areas, estimated damage costs differed significantly between SCE_ 2, 3 and 4. The use of a DEM with higher resolution (2 m resolution) in Samut Sakhon city would be one factor supporting a more accurate determination of the extent of inundated areas (and flood depths).

5 Conclusion

This research developed a methodology for assessing climate change risk for urban roads at the local level. Adopting RVA approaches and using NK-GIAS for more detailed analyses resulted in a better understanding of potential hazards, vulnerabilities and adaptive capacities. The six cities selected for this study were judged to be susceptible in a variety of ways. The cities were particularly vulnerable to floods and saline intrusion due to their geographic location and the threat from SLR and storm surge. Road infrastructure was identified as being most susceptible to the impacts of climate change, though current policies related to UTI do not address climate change.

This study shows how economic losses can be linked to various flood scenarios, with increasing depth directly increasing damage costs in each city. The estimated maximum road damage costs for Hoi An, Kampot and Samut Sakhon were 0.95, 1.5 and 6.5 million USD, respectively, based on the results of the NK-GIAS analysis under scenario 1. Under SCE 4 (the most severe scenario) the estimated maximum road damage costs are 20 million USD for Hoi Anh, 3 million USD for Kampot and 21 million USD for Samut Sakhon. However, once a dike was built in Samut Sakhon in 2009, and flooding was decreased in scenarios SCE 1 to SCE 3 (water levels were lower than the height of the dike), the real damage expenses were greatly reduced. However, using SCE_4, the maximum damage road cost will be 21 million USD.

Two key limitations were identified in the assessment methodology. The use of a DEM with low resolution, which preferably should not be larger than 5 m for the input DEM, may reduce accuracy in the flooded area. Assessment of flood damage should use various techniques for examining all vulnerability issues and each city's susceptibility to flooding.

Local governments have developed engineered adaptation strategies to deal with flooding, including the construction of dikes and embankments for long-term protection and the use of sandbags as temporary solutions. Samut Sakhon, where the frequency of floods was reduced by more than 70% after dike construction surrounding the city was completed, is an example of how efficient dike protection can be (Samut Sakhon Municipality 2018). However, more recent considerations have focused on the need to factor climate change into planning and policy. Regarding UTI, especially the accompanying design and building criteria, the current climate change action plans (through 2050) for Vinh Long and Hoi An

(Vietnam) are very generic. While not discussed in the present chapter, it would be a fascinating follow-up. For future work, it would be useful to follow up with stakeholders to discuss the extent to which climate change is being considered in UTI developments.

Based on the study's findings, several recommendations are made: (i) Capacity building should especially be offered to local government employees in the cities of Hoi An, Kampot and Samut Sakhon to promote the mainstreaming of climate change into policy and planning; (ii) To ensure that UTI development is affordable and climate resilient, encourage collaboration among the private sector, city governments and local government; (iii) Use the best available data and information (historical and from modelling) on water levels to inform infrastructure design; (iv) if research had been concentrated on creating new damage curves in the laboratory using physical models to mimic damage due to both flood depth and duration, the results of the road damage assessment might have been improved. Further research is recommended to do more research and laboratory tests to improve road damage curves for the NK-GIAS analysis.

Acknowledgements We acknowledge and appreciate the support of the Asia-Pacific Network for Global Change Research (APN) for providing the research grant and guidance, especially from Dr Linda Anne Stevenson, Ms Dyota Condrorini, Dr Nafesa Ismail and Ms Christmas Uchiyama; the technical support from Nippon Koei Co., Ltd—Japan, especially the support from Mr Shigeru Nakamura in providing free access to NK-GIAS software; the Southern Institute of Water Resource Research, HCMC, Vietnam; the Transport Sustainable Development and Environmental Research Institute, Hanoi, Vietnam; Research Institute of Transportation, Agriculture and Environment, Hanoi, Vietnam; Royal University of Phnom Penh, Cambodia; Southeast Asia START Regional Centre, Chulalongkorn University, Bangkok, Thailand; and Green Style Co., Ltd, Bangkok, Thailand.

References

ADB. 2009. The Economics of Climate Change in Southeast Asia: A Regional Review. Manila, Philippines.

ADB. 2010. Ho Chi Minh City, Adaptation to Climate Change—Summary Report. Manila, Philippines.

De Bruijn. 2005. Resilience and Flood Risk Management. In *A Systems Approach Applied to Lowland Rivers*.

Füssel, Hans-Martin, and Richard J. T. Klein. 2006. "Climate Change Vulnerability Assessments: An Evolution of Conceptual Thinking." *Climatic Change* 75 (3):301–329. https://doi.org/10.1007/s10584-006-0329-3.

Green Style Co., Ltd. 2018. Climate Change Risk Assessment for Flooding—Loss and Damage of Roads in Samut Sakhon City (internal project report). Bangkok, Thailand.

Hens, Luc, Nguyen An Thinh, Tran Hong Hanh, Ngo Sy Cuong, Tran Dinh Lan, Nguyen Van Thanh, and Dang Thanh Le. 2018. "Sea-Level Rise and Resilience in Vietnam and the Asia-Pacific: A Synthesis." *Vietnam Journal of Earth Sciences* 40 (2):126–152.

Hoi An PC. 2015. Socio-Economic Development in Hoi An (internal report). Hoi An, Vietnam.

Hoi An PC. 2022. Socio-economic Development in Hoi An (internal report). Hoi An, Vietnam.

Hoi An Transportation Department. 2017. Report of Transportation Infrastructure in Hoi An (internal report). Hoi An, Vietnam.

Hua Hin Municipality. 2015. "Hua Hin Municipality", accessed October 2003. http://www.huahin.go.th/new/content/general.

Intergovernmental Panel on Climate, Change. 2014. *Climate Change 2014—Impacts, Adaptation and Vulnerability: Part B: Regional Aspects: Working Group II Contribution to the IPCC Fifth Assessment Report: Volume 2: Regional Aspects*. Vol. 2. Cambridge: Cambridge University Press.

IPCC. 2007. Climate Change 2007. Impacts, Adaptation and Vulnerability. Contribution of Working Group II to the Fourth Assessment Report of the Intergovernmental Panel on Climate Change, edited by M.L. Parry, Canziani, O.F., Palutikof, J.P., van der Linden P.J., & Hanson, C.E. Cambridge University Press, Cambridge, UK.

IPCC. 2013. Climate Change 2013: The Physical Science Basis. Contribution of Working Group I to the Fifth Assessment Report of the Intergovernmental Panel on Climate Change, edited by T.F. Stocker, Qin, D., Plattner, G.-K., Tignor, M., Allen, S.K., Boschung, J., Nauels, A., Xia, Y., Bex, V., & Midgley, P.M. . Cambridge University Press, Cambridge, UK and New York, NY, USA.

IPCC. 2014. Climate Change 2014: Impacts, Adaptation, and Vulnerability. Part B: Regional Aspects. Contribution of Working Group II to the Fifth Assessment Report of the Intergovernmental Panel on Climate Change, edited by V.R. Barros, Field, C.B., Dokken, D.J., Mastrandrea, M.D., Mach, K.J., Bilir, T.E., Chatterjee, M., Ebi, K.L., Estrada, Y.O., Genova , R.C., Girma, B., Kissel, E.S., Levy, A.N., MacCracken, S., Mastrandrea, P.R., & White, L.L. . Cambridge University Press, Cambridge, United Kingdom and New York, NY, USA.

ISDR. 2002. Disaster Reduction and Sustainable Development Understanding the links between vulnerability and risk to disasters related to development and environment. https://www.gdrc.org/uem/disasters/disenvi/DR-and-SD-English.pdf. International Strategy for Disaster Reduction, United Nation.

JICA. 2013. Project for the Comprehensive Flood Management Plan for the Chao Phraya River Basin. Japan International Cooperation Agency (JICA). In *Final Report Volume 1: Summary Report, September 2013*.

JICA. 2015. Preparatory Survey Report on the Project for Expansion and Improvement of Water Supply System in Kampot and Shihanouk Village in the Kingdom of Cambodia. Kitakyushu City, Japan.

Jorik Chen. 2007. "Flood Damage Map for the Huong River Basin." Bachelor, Civil Engineering, University of Twente, The Netherlands.

Kampot Administration. 2015. Report of Transportation Infrastructure in Kampot (internal report). Kampot, Cambodia.

MPWT. 2013. Climate Change Strategic Plan for Climate Change Adaptation and Greenhouse Gas Mitigation—Transport Sector. Ministry of Public Work and Transports (MPWT), Phnom Penh, Cambodia.

Noi, Lam Vu Thanh, and Vilas Nitivattananon. 2015. "Assessment of Vulnerabilities to Climate Change for Urban Water and Wastewater Infrastructure Management: Case Study in Dong Nai River Basin, Vietnam." *Environmental Development* 16:119–137.https://doi.org/10.1016/j.envdev.2015.06.014

Reisinger, Andy, Mark Howden, Carolina Vera, Mathias Garschagen, Margot Hurlbert, Sylvia Kreibiehl, Katharine J. Mach, Katja Mintenbeck, Brian O'Neill, Minal Pathak, Roque Pedace, Hans-Otto Pörtner, Elvira Poloczanska, Maisa Rojas Corradi, Jana Sillmann, Maarten van Aalst, David Viner, Richard Jones, Alexander C. Ruane, and Roshanka Ranasinghe. 2020. The Concept of Risk in the IPCC Sixth Assessment Report: A Summary of Cross-Working Group Discussions. Intergovernmental Panel on Climate Change, Geneva, Switzerland.

RUPP. 2018. Climate Change Risk Assessment for Flooding—Loss and Damage of Roads in Kampot city (internal project report). Phnom Penh, Cambodia.

Samarakon L., Nakamura, S., & Hunukumbura, P. 2012. Geographic Information System (GIS) using NK-GIAS. Nippon Koei Co., Ltd, Japan.

Samut Sakhon Municipality. 2018. Report of Transportation Infrastructure in Samut Sakhon (internal report), Kampot, Thailand.

SIHYMECC. 2015. Research on Mapping Flooding in Vietnam (project report), HCMC, Vietnam.

Un Habitat. 2011. Global Report on Human Settlements 2011: Cities and Climate Change. Earthscan.

Vietnam Ministry of Construction. 2015. Decison No 1161/QD-BXD dated 15 October 2015 about Announcement of Investment Capital for Building Construction and Working Construction Projects in 2014. Vietnam.

Vinh Long PC. 2015. Socio-economic Development in Vinh Long (internal report), Vinh Long, Vietnam.

Vinh Long PC. 2022. Socio-Economic Development in Vinh Long (internal report), Vinh Long, Vietnam.

World Bank. 2010. The Social Dimensions of Adaptation to Climate Change in Vietnam. Economics of Adaptation to Climate Change. Discussion Paper Number 12, Nov 2010.

CHAPTER 14

Forward-Thinking Perspectives Towards Integrating Climate Change Adaptation and Disaster Risk Reduction for Synergetic Resilience-building and Addressing Loss and Damage

Dennis Mombauer and Vositha Wijenayake

Abstract Climate change adaptation (CCA) and disaster risk reduction (DRR) are deeply interlinked and aim to achieve shared objectives of resilience-building and averting, minimizing, and addressing loss and damage (L&D). Current global frameworks such as the Paris Agreement or the Sendai Framework recognize these linkages, which have significant implications in terms of how CCA and DRR are planned and implemented. However, implementation processes on all levels often do

D. Mombauer (✉) · V. Wijenayake
SLYCAN Trust, Colombo, Sri Lanka
e-mail: dennis@slycantrust.org

V. Wijenayake
e-mail: vositha@slycantrust.org

V. Wijenayake et al. (eds.), *Linking Climate Change Adaptation, Disaster Risk Reduction, and Loss & Damage*,
https://doi.org/10.1007/978-981-99-8055-0_13

not transform these linkages into operational advantages. Country-level experiences and learnings can offer entry points to identify potential interconnections, overlaps, co-benefits, and synergies to enhance resilience-building and better address L&D. Based on the case studies in this publication, this forward-thinking paper explores what kinds of evidence, information, institutional arrangements, and monitoring frameworks could be needed to enhance and operationalize synergies and develop a holistic framework that links CCA and DRR to effectively build resilience and address L&D.

Keywords Adaptation · Disaster risk reduction · Loss and damage · Resilience-building · Forward-thinking

1 Introduction

Climate change adaptation (CCA) and disaster risk reduction (DRR) are interlinked and aim to achieve similar objectives: building resilience and averting, minimizing, and addressing loss and damage (L&D). Global frameworks such as the Paris Agreement (PA) and the Sendai Framework for Disaster Risk Reduction (SFDRR) recognize these linkages and provide avenues for creating synergies and possible pathways towards a holistic resilience-building approach in the light of increasing climate-induced L&D and a complex landscape climate- and non-climate-related risks.

CCA, as defined by the Intergovernmental Panel on Climate Change (IPCC AR6 WGII), is "the process of adjustment in human or natural systems to actual or expected climate and its effects, in order to moderate harm or exploit beneficial opportunities." It is often "organized around resilience," which broadly describes "the ability to maintain essential function, identity, and structure" of a system as well as its "capacity for transformation."

DRR, as defined by the United Nations Office for Disaster Risk Reduction (UNDRR) in the context of the SFDRR, means activities "aimed at preventing new and reducing existing disaster risk and managing residual risk, all of which contribute to strengthening resilience and therefore to the achievement of sustainable development." Under the SFDRR, the global goal for DRR is "the substantial reduction of disaster risk and losses in lives, livelihoods, and health and in the economic, physical, social, cultural, and environmental assets of persons, businesses, communities and countries."

While L&D in the context of climate change does not have a universally used exact definition, it broadly describes the residual impacts of climate change that could not or have not been averted, i.e., unavoidable or unavoided impacts that exceed the soft or hard limits of adaptation and climate-related risk reduction. Accordingly, both CCA and DRR aim to minimize, avert, and address L&D, opening the possibility of identifying and harnessing co-benefits, synergies, and positive reinforcement between both areas and their associated institutional setups. As highlighted in the IPCC's Sixth Assessment Report, "climate risk and disaster risk are increasingly addressed together, bridging the climate change adaptation and disaster risk reduction communities."

Explicit and implicit references to this interconnected nature of CCA and DRR can be found in the respective intergovernmental processes and treaties that govern CCA and DRR. For example, Article 8 of the Paris Agreement outlines "areas of cooperation and facilitation to enhance understanding, action, and support" with respect to L&D associated with the adverse effects of climate change. This includes areas linked to CCA as well as to DRR, including early warning systems; emergency preparedness; comprehensive risk assessment and management; risk insurance facilities, climate risk pooling, and other insurance solutions; and building the resilience of communities, livelihoods, and ecosystems.

Similarly, the SFDRR as the primary global framework for DRR makes a clear reference to disasters being "exacerbated by climate change and increasing in frequency and intensity, significantly [impeding] progress towards sustainable development."

These theoretical and practical linkages have significant implications in terms of how CCA and DRR are planned and implemented (Hoang et al., 2023). However, in many cases, these linkages are not translated into operational advantages on the national and local level, especially for addressing climate-induced L&D. As the IPCC AR6 report outlines, "linking disaster risk reduction and climate change adaptation can also be an important basis for discussion in climate negotiations regarding the allocation of funds needed for tackling climate change," which is especially important in the context of the process of operationalizing the new L&D fund and funding arrangements.

Past progress in CCA and DRR provides us a view on how future actions can be moderated. This publication contains examples and lessons learnt regarding the integration of CCA and DRR for inclusive and scaled-up resilience of communities as well as national-level policy processes interlinked with global processes such as those of the UNFCCC and the SFDRR, illustrating the possibility of creating holistic and impactful policy initiatives through integrated approaches and enhanced multi-stakeholder coordination.

2 The Interface(s) Between CCA, DRR, and L&D

As outlined above, there are many similarities between DRR and CCA, particularly in view of their shared goal of resilience-building and averting, minimizing, and addressing L&D to protect vulnerable households, communities, infrastructure, and systems. However, there are also significant differences between the two approaches, such as a different overall scope, different spatial and temporal scales, different institutional frameworks, and different indicators, evaluation criteria, knowledge bases, and data formats.

2.1 Interfaces and Barriers for Integration

The common goal of building resilience and averting, minimizing, and addressing L&D provides a strong teleological foundation to connect CCA and DRR. As the IPCC AR6 report highlights, "the concepts of adaptation, vulnerability, resilience, and risk provide overlapping, alternative entry points for the climate change challenge." Regarding resilience and L&D, a core concept and connection point is "vulnerability," which is "a component of risk, but also an important focus independently, improving understanding of the differential impacts of climate change on people of different gender, race, wealth, social status, and other attributes. Vulnerability also provides an important link between climate adaptation and disaster risk reduction [...] [as well as] the climate risk and disaster risk communities, recognizing complementarities and differences between these communities" (IPCC, 2022).

Building on this, three major pathways to strengthen the interconnection between CCA and DRR could be outlined as follows:

- Identifying components and prerequisites for resilience-building and mapping out the contributions of CCA and DRR practices to identify complementarities or gaps;
- Focusing on vulnerability of individuals, households, communities, or countries to identify how CCA and DRR are currently reducing it on different timescales and with regard to different risks;
- Assessing L&D caused by climate change and understanding why it has not been averted, minimized, or addressed through either CCA or DRR, thereby locating gaps and needs for enhancement through a synergetic mechanism.

In general, DRR is often more focused on the local level and short timeframes, with plans designed by or in close collaboration with local authorities under the auspices of national agencies for disaster management, public security, defence, or internal affairs, covering climatic as well as non-climatic disasters (such as earthquakes). Conversely, CCA tends to be planned at the national level, often under the Ministry of Environment, and targets medium to long timeframes that include slow-onset processes and increasing weather variability in addition to extreme events and disasters.

There is a need to better understand overlaps, co-benefits, and potential synergies between CCA and DRR for addressing L&D. It is particularly important to focus on the national-level picture on how CCA and DRR are interlinked in the context of L&D, mapping tensions and trade-offs. In this regard, the lack of robust and universally accepted methodologies for assessing and quantifying L&D poses a challenge towards addressing climate impacts in an effective manner.

2.2 Global Policy Frameworks and Processes

Many policy frameworks at the global level offer entry points for operationalizing interlinkages and synergies between climate change, resilience-building, DRR, and averting and minimizing L&D. Among these are the UNFCCC, the Paris Agreement, the Agenda 2030 for Sustainable Development, the SFDRR, and other processes aimed at climate and disaster risk management, such as the InsuResilience Global Partnership or the Global Shield Against Climate Risks.

Under the UNFCCC and the Paris Agreement, climate change adaptation is a key focus which has seen wide growth and evolved to address a multitude of thematic areas over the years, including broader approaches to addressing resilience-building and climate risk reduction. This includes processes such as those around the National Adaptation Plans (NAPs) and the Nationally Determined Contributions (NDCs) under the Paris Agreement. Additionally, with the progress of climate science, reporting related to vulnerability and adaptation measures through national communications has progressed to include climate risk assessments, which take into consideration indicators focused on risk management, including disaster risk management. Further, the NDCs submitted by countries focus on resilience-building as a key component under climate action, which create avenues for integrating DRR and CCA.

The UNFCCC process and its different thematic tracks provide opportunities for the inclusion of risk management. For example, this includes the Cancun Adaptation Framework (FCCC/CP/2010/7/Add.1) which established several key adaptation-related processes and bodies such as the Adaptation Committee and provided a framework for long-term cooperative action. Further, under its actions related to adaptation, the Cancun Framework requests all Parties to take actions which will contribute to enhancing action on adaptation, among which is the request for enhancing of DRR strategies that are relevant to action on climate change. The framework also has risk reduction as one of its key components, which interlinks closely with long-term climate resilience and planning processes related to DRR.

In addition to the above, the framework also established the process for NAPs focused on enabling Least Developed Country (LDCs) Parties to work on NAPs for national-level adaptation planning, including formulating and implementing adaptation plans, as well as identifying adaptation needs to be addressed in developing countries. Today, the process remains closely connected to financial institutions such as the Green Climate Fund, which was established under the same framework and through which funding is accessed by countries for risk management with the aim of reducing climate vulnerabilities. It is important to note that climate and disaster risk management as well as risk assessment form key components of the NAP process. Out of the 53 NAPs submitted by developing country Parties to NAP Central so far, the majority of Parties refer to DRR as a key action point in their NAP or as an entry point for building climate resilience through interlinkages between DRR and climate action.

Further processes which interlink DRR and climate risk management include the activities under the Warsaw International Mechanism for Loss and Damage associated with Climate Change Impacts (WIM), which focuses on risk management as one of its key work areas. This includes risk management through early warning and preventive measures that would converge with adaptation as well as risk transfer and risk retention which would move further into the area of L&D-related work. Other aspects that remain synergized include common areas of work related to human mobility, resettlement, and relocation, which highlight common thematic focus areas related to the work under the Sendai Framework and DRR, as well as activities related to adaptation and L&D under the UNFCCC process.

The Paris Agreement also provides key entry points for the integration of DRR, CCA, and L&D. Some potential avenues for integration include the processes related to the Global Stocktake (GST), where CCA is an important component, and the Global Goal on Adaptation (GGA), which would provide opportunities for identifying key indicators and means for achieving necessary action.

In addition to the UNFCCC process, the Agenda 2030 for Sustainable Development forges key interlinkages between disaster management, climate action, and other components related to long-term and sustainable resilience-building. Since 1994, the interlink between DRR and sustainable development has been laid out, and strong proponents for this include the Yokohama Strategy and Plan of Action for a Safer World. Under the 2030 Agenda for Sustainable Development, SDG 11 on sustainable cities and communities has DRR as a key component including policy frameworks interlinked with disaster risk management. Additionally, SDG 13 on climate action also provides an entry point for harnessing synergies between global, regional, and national climate action to be integrated into the SDG process, seeking to create common approaches across processes to achieve scaled-up resilience, which is already reflected in most submitted NAPs.

Furthermore, it is also important to observe processes focused on providing finance and other means of implementation with regard to climate and disaster risk finance and insurance (CDRFI). For example, the newly established Global Shield aims to provide finance for building resilience to disasters and climate risks, which could provide opportunities for synergies through common actions aimed at building the long-term resilience of vulnerable communities to both climate and disaster risk through evidence- and science-based approaches.

3 Learning from the Case Studies

Based on the case studies contained in this publication as well as other available evidence from the Asia–Pacific region, there are several key findings in relation to resilience-building through CCA and DRR as well as addressing L&D. The case studies highlight several main causes and forms of L&D, key impacted sectors, and existing methodologies and tools to assess, understand, measure, and/or quantify L&D.

3.1 Understanding L&D in the Context of Climate Change

Agriculture, rural livelihoods, energy, infrastructure, and natural resources are among the priority sectors in the Asia–Pacific region which are exposed to severe, complex, cascading, and often covariate risks that are caused by or related to climate change. This includes climate variability leading to increasing uncertainty, climate-related extreme events and disasters, and long-term climatic changes to average conditions and risks (van der Geest, 2023).

There is a general distinction between economic and non-economic L&D, which is often defined in relation to goods and services that are commonly traded on markets. In this sense, economic losses include production losses, decline in productivity, loss of assets, affected structures (such as houses, roads, hospitals, or schools), and loss of wealth, but also costs for rehabilitation and reconstruction (Lotia, 2023). Non-economic losses, on the other hand, describe impacts to culture, social cohesion, tangible or intangible heritage, natural ecosystems, mental and psychosocial health, which are more difficult to valuate and often cannot be readily quantified in monetary terms.

Both kinds of L&D are interconnected, often without clear boundaries between each other, and can cause indirect or cascading secondary impacts, such as changes to local labour market demand, prices and scarcity of goods, or mental health impacts. In many cases, L&D erodes the resilience of communities and heightens their vulnerability to future shocks and impacts in a multitude of ways. For example, a flood can affect livestock health and feed availability, forcing farmers to sell at below market rates if they lack the capital to treat the animals and replace feed.

In addition to lost income, this also represents a loss of financial assets serving as buffers against future shocks, which cannot be easily replaced. Similarly, pests, waterborne diseases, or crop failures are more likely to occur in the aftermath of floods due to changes in soil biotic conditions and waterlogging of areas (Lotia, 2023).

Furthermore, vulnerability, resilience, and the effectiveness of CCA or DRR measures is highly context-specific and depends on demographic characteristics, income, wealth, education, willingness to take risks (associated with higher life satisfaction), and a multitude of other factors (Basnayake et al., 2023).

3.2 Approaches Towards Linking CCA and DRR

Effectively linking CCA and DRR requires a robust understanding of the determinants and drivers of both resilience and L&D. These could be related to spheres of private and economic life, for example, agriculture, livestock, MSMEs, human health, or human mobility (Lotia, 2023); available capacities, coping strategies, and preventive measures, such as selling assets, alternative income sources, social networks, insurance, physical protection, food aid, or relief (van der Geest, 2023); or the different institutional structures and policy processes under the respective global frameworks.

There are different categories of interventions and measures that can respond to and address different forms of L&D, many of which are contained in the case studies in this publication. For economic L&D, structured or unstructured actions include planned relocation, assisted migration, reskilling, alternative livelihood provisions, compensation, social protection, risk transfer, recovery, rehabilitation, stimulus, reconstruction, shelter and housing, compensation, microloans, community funds, or the development of professional associations (Nhat and Thinh, 2023; Boyland et al., 2023).

Similarly, for non-economic L&D, measures could include recognition and repair of loss, enabling safe access to abandoned sites, active remembrance, counselling, official apologies, ecosystem restoration or rehabilitation, memorialization, documentation, healthcare, or social assistance (Nhat and Thinh, 2023; Boyland et al., 2023; Mombauer and Wijenayake, 2023).

As this is a wide range of potential actions that cover a multitude of levels—from local to national or regional—and sectors—such as health, education, finance, labour, social protection, infrastructure, human settlements—the available options for addressing L&D blur the distinctions between CCA and DRR and often could be classified as either depending on the exact circumstances. Bringing them together into a holistic framework presents a serious challenge, particularly regarding the existing institutional setups, policy environments, funding streams, and methodologies.

3.3 Towards an Integrated National Mechanism

Several of the papers outline potential functions and the shape of a national mechanism for L&D, which could bring together and coordinate CCA and DRR activities to address L&D in a holistic way. Such a mechanism could strengthen institutional coordination, knowledge-sharing, vertical and horizontal integration, M&E, MRV, budget and progress tracking, L&D assessments, and stakeholder engagement. It could also improve knowledge and understanding of global risk management approaches and L&D assessment methods, mobilize additional financial means of implementation, and bridge the gaps between economic and non-economic L&D as well as between disaster events and slow-onset processes (Prabhakar, 2023b).

Integrated actions at national and local level could contribute to advancing actions towards long-term resilience-building of vulnerable communities and ecosystems. This could be through activities at national and subnational level related to NAPs, where readiness support funding via the GCF could be an opportunity to create national-level synergy for planning to integrate DRR and CCA. Additionally, through the NAP readiness actions, there could be policy coherence developed between policy and action related to DRR and CCA. From the assessments conducted for accessing funding to readiness activities that focus on holistic resilience of communities and ecosystems through Readiness Funding, the NAP process could provide entry points to create interlinkages among several policy processes at national level, including DRR, CCA, SDGs, as well as L&D with regard to risk management.

Additionally, climate commitments under the Paris Agreement also provide important opportunities for integration of DRR, CCA, and other processes. For example, NDCs of countries provide commitments at national level for policy coherence. In Sri Lanka, the updated NDCs submitted in 2021 indicate a focus on long-term resilience-building;

achieving sustainable development through several action points which are focuses at sectoral level; and interlinkages and synergies regarding L&D through commitments on addressing climate-induced L&D, which are closely connected to DRR action in the country, through coordination processes which are interlinked to the disaster management as well as overlapping of common priorities such as climate and disaster risk transfer in the agriculture sector in the country.

4 Operationalizing Synergies between CCA and DRR

Any attempt to operationalize the synergies between the existing frameworks and current as well as potential actions must consider at least three key areas: (a) evidence and information, (b) institutional arrangements, and (c) monitoring, evaluation, and reporting systems.

4.1 Evidence and Information

When it comes to data collection, understanding the different perspectives, priorities, and existing systems, for example, from a DRR or disaster management perspective, evaluations are primarily carried out to estimate losses and damages for statistical purposes as well as for reconstruction and rehabilitation. However, from a CCA perspective, the evaluation or assessment is mostly done to estimate loss and damage with the goal of minimizing the impacts of future events (Basnayake et al., 2023).

Furthermore, DRR and CCA often use common or similar indicators, but collect data in separate processes and with different methodologies. In particular, it is important to understand the determining factors for vulnerability, the needs of vulnerable populations, and the non-economic and indirect forms of L&D. Assessments should cover not only the nature of assets that are lost, but also the reason for and manner of their loss, as this could provide key insights towards identifying adaptation limits and constraints (van der Geest, 2023).

Holistic evidence collection that bridges the gaps between DRR and CCA needs to use locally appropriate indicators that are identified in a consultative manner (Prabhakar, 2023a) while also including metrics for non-economic L&D, such as indicators for psychological wellbeing or life satisfaction data on the individual level (Basnayake et al., 2023).

4.2 Institutional Arrangements

As outlined earlier, DRR and CCA are addressed through different institutional structures and policy processes. Some potential entry points to connect them across shared dimensions of vulnerability, resilience, and L&D include the following:

- Linking NDC implementation and reporting on the subnational level with SDG localization processes and subnational implementation under the SFDRR (Mombauer and Wijenayake, 2023).
- Aligning data collection and data-sharing mechanisms under the SFDRR and the SDGs with NDC review and implementation processes (Mombauer and Wijenayake, 2023).
- Facilitating self-organization of affected communities to strengthen social safety nets, early warning, planning processes, and targeted assistance (Boyland et al., 2023).
- Identifying co-benefits for accessing finance and support, including for capacity-building and awareness creation measures (Mombauer and Wijenayake, 2023).
- Strengthening devolved, participatory, and cross-level decision-making and multi-agency and multi-scale inclusion to avoid duplication of efforts (Boyland et al., 2023).
- Operationalizing "building back better" principles and developing clear guidelines, milestones, and agency responsibilities for transitioning from recovery to development (Boyland et al., 2023).
- Setting up a dedicated institutional framework for MRV to coordinate pre-existing data collection mechanisms and align them with reporting under the Paris Agreement, Agenda 2030, and the SFDRR (Mombauer and Wijenayake, 2023).

4.3 Monitoring, Evaluation, and Reporting Systems

Finally, any evidence collected and activities managed by the relevant institutions need to be regularly assessed, monitored, evaluated, reported on, and improved. In order to operationalize synergies, it is crucial to develop and consolidate monitoring and evaluation systems for the regular assessment of risks and vulnerabilities, which allows short-term goals (such as those related to concrete DRR or CCA measures) to be turned into long-term strategies of resilience-building that utilize continuous learning mechanisms.

This would also allow countries and other stakeholders to identify DRR activities that build adaptive capacities and resilience through a detailed analysis of monitoring data and the vulnerability assessments, as well as other monitoring and evaluation activities (Nhat and Thinh, 2023).

5 Conclusion

A pivotal question for global, regional, national, and local action over the next decades will be how to move towards a holistic framework that links CCA and DRR to address L&D effectively and comprehensively and operationalize synergies. What are the key challenges and issues to achieve this, and what solutions could it present?

While all countries would have these different processes, there is a need to better integrate the processes at national level through institutional capacity-building as well as coordination mechanisms being effectively functioned. This includes the establishment of processes for coordination among ministries and institutional structures as well as multiple stakeholders focused on CCA and DRR across national and global processes.

To achieve this, it is vital that there is a good understanding of global processes related to the different processes aimed at building long-term resilience, as well as sufficient evidence, knowledge, and technical expertise which would allow for countries and stakeholders to ensure that those who are most vulnerable to climate and disaster risks are benefitted through practical and viable solutions as well as measures leading to sustained and sustainable development.

Identifying approaches for creating synergies considering community needs adaption needs, and the need to avert, minimize, and address climate-induced L&D while achieving SDGs require long-term vision and strategic interventions which look inside and outside of the existing global processes. Two examples of such approaches could be the Loss and Damage Fund under the UNFCCC and the Global Shield against Climate Risks, which aim to provide finance for the same purpose (addressing climate-induced L&D) but with a different functionality. It is important to understand these dualities to be able to benefit from existing structures to achieve the common objectives pertaining to policy and action, which will be to ensure long-term climate and disaster resilience of all, through synergized and collective approaches which avoid duplicating efforts and drawing the maximum use of available means of implementation to address the urgent needs of climate-vulnerable populations at all levels.

References

Basnayake, Senaka, Guota, Niladri, Rahman, Md. Habibur, Jayasinghe, Susantha, Ulubasoglu, Mehmet Ali, & Tong, Lan Anh (2023). Flood-induced losses and damages to agricultural production and farmers' mental wellbeing in Sri Lanka.

Boyland, Michael, Marks, Danny, Kimkong, Ham, & Nugroho, Agus (2023). Towards effective loss and damage systems in disaster recovery contexts in Southeast Asia.

Hoang, Hop T. B., Lasco, Rodel, Nguyen, Ninh H., & Boer, Rizaldi (2023). SWOC analysis on the proposed linkage between Climate Change Adaptation (CCA), Disaster Risk Reduction (DRR), and Loss and Damage (L&D): Case Studies in the low-lying coastal cities of Indonesia, Philippines, Thailand and Vietnam.

Intergovernmental Panel on Climate Change (IPCC). (2022). Climate Change 2022. Impacts, Adaptation and Vulnerability. Working Group II contribution to the Sixth Assessment Report of the Intergovernmental Panel on Climate Change.

Lotia, Hina (2023). Pillars of resilience: Assessing loss and damage at the local level.

Mombauer, Dennis & Wijenayake, Vositha (2023). Policy gaps and needs analysis for the implementation of NDCs on adaptation and loss and damage in Bangladesh, Nepal, and Sri Lanka.

Nhat, Le Minh & Thinh, Dang Quang (2023). Linkages between disaster risk reduction and climate change adaptation in the context of increasing climate change-induced loss and damage in Vietnam.

Prabhakar, S. V. R. K., Chiba, Yohei, & Islam, Md. Atikul (2023a). Identification of non-economic loss and damage (NELD) indicators and practices in the context of climatic events.

Prabhakar, S. V. R. K., Pereira, Joy Jacequline, Pulhin, Juan M., Thang, Nguyen Van, Nyda, Chhinh, Aquino, Dannica Rose G., Nga, Pham Thi Thanh, Mau, Nguyen Dang, Thinh, Dang Quang, & Thuy, Tran Thanh (2023b). Priorities for addressing Slow-Onset Events (SOEs) in selected ASEAN countries.

Singh, Harjeet & Sharma, Sindra (2023). A people-centred approach to assess and address impacts of climate change induced loss & damage.

Van der Geest, Kees (2023). A conceptual framework and research design for assessing losses and damages from climate change in vulnerable communities.

Index

V. Wijenayake et al. (eds.), *Linking Climate Change Adaptation, Disaster Risk Reduction, and Loss & Damage*, https://doi.org/10.1007/978-981-99-8055-0

SPRINGER NATURE

GPSR Compliance

The European Union's (EU) General Product Safety Regulation (GPSR) is a set of rules that requires consumer products to be safe and our obligations to ensure this.

If you have any concerns about our products, you can contact us on ProductSafety@springernature.com

In case Publisher is established outside the EU, the EU authorized representative is:

Springer Nature Customer Service Center GmbH
Europaplatz 3
69115 Heidelberg, Germany

The manufacturer's authorised representative in the EU is Springer Nature Customer Service Centre GmbH, Europaplatz 3, 69115 Heidelberg, Germany. If you have any concerns regarding our products, please contact ProductSafety@springernature.com

Printed and bound by CPI Group (UK) Ltd, Croydon, CR0 4YY
01/10/2026
02225924-0007